THE HEALING POWER OF YOUR MIND

The Doctor Alone Can't Cure You

by

Rolf Alexander, M.D.

Healing Arts Press
Rochester, Vermont

Healing Arts Press
One Park Street
Rochester, Vermont 05767

LIBRARY OF CONGRESS CATALOGING IN PUBLICATION DATA
Alexander, Rolf, b. 1891.
[Renewing power of your mind]
The healing power of your mind : the doctor alone can't cure you / Rolf Alexander.
p. cm.
Reprint. Originally published: The renewing power of your mind. New York, N.Y. : Warner Books, c1976.
ISBN 0-89281-339-3
1. Mental healing. I. Title.
RZ400.A76 1989 89-17176
616'.0019—dc20 CIP

Printed and bound in the United States.

10 9 8 7 6 5 4 3 2 1

Healing Arts Press is a division of Inner Traditions International, Ltd.

Distributed to the book trade in Canada by Book Center, Inc., Montreal, Quebec
Distributed to the health food trade in Canada by Alive Books, Toronto and Vancouver

Contents

Publisher's Preface

Much of the material in this book appeared as a privately published handbook for the author's own patients. It is no exaggeration to say that it contributed greatly to the recovery of many who previously had looked upon themselves as "incurable" and that it shortened the duration of the illness to a considerable degree in every case where its principles were followed persistently.

This book was not written for entertainment purposes; it contains specific information that must be mastered by mental effort if you would reap benefits. The principles and the exercises outlined herein have been tested over a period of years by actual practice, and their effects on hundreds of patients have been checked and double-checked. These practices, wedded as they are to the practices of Tibetan and Indian healing methods, will transform your life and lead you to self-mastery through "The Healing Power of the Mind."

1

The Quest

"We take the first step toward knowledge, my son, when we realize our ignorance. The difference between the discussion heard in the Almshouse and that heard in the Lama's council room, is that the former is by deluded men who believe that THEY are wise, while the latter, realizing their ignorance, draw wisdom from the gods."

Book of Right Feeling

The practice of medicine has been longstanding in my family. I, of course, wanted to carry on the tradition and so, more than a half century ago, I graduated from medical school—with honors—certain that I could be a successful healer.

As weeks grew into months, and as case after case passed through my hands for treatment, a sense of frustration and disillusionment began to grow within me as I came to realize that the therapeutics I had mastered at such expense of time and effort either worked indifferently or not at all.

In fact, I began to suspect that this profession of medicine, far from being the miracle-working science I had believed it to be, might instead be an assemblage of unproven and *unprovable* theories organized around a few basic facts.

I began asking questions but found no satisfactory answers, even from the most eminent of my professional acquaintances.

"Why was a treatment sometimes effective in one case, while in another, perhaps one suffering from the same disease in a milder form, it was utterly impotent?"

"Why did the same drugs have such a radically different effect on different people?"

"Why did some people die from comparatively minor causes while others, literally riddled with disease, hung on to life for years?"

"Was there a deeper, more fundamental factor which governed the human organism and determined its susceptibility to disease and its response to treatment? Was disease the effect of a failure of this unknown factor to function?"

The more I thought over the latter question, the more the experiences of my daily practice seemed to confirm that there was indeed a potent basic factor within humankind which ordered and regulated all of the body's vital functions, attended to all of the body's defense mechanisms, ordered their repair of damaged bones and organs and tissues, and which tied all of the numerous activities within the body into a smoothly functioning whole.

From that point of view, the treatment of disease became merely the elimination of effects, without necessarily altering the causes. True, this battle against the effects called disease was a necessary battle, but without touching that deeper unknown cause, the best that could be hoped for would be the mere postponement of death, since inevitably the leaking waters of life would break the crumbling body structure at another place.

In search of this basic factor within humankind, I spent the next few years at post-graduate work in London, Paris, Vienna and Heidelberg, where I sat at the feet of some of the most famous pathologists, endocrinologists and analytical psychologists of that day. And though the work of these great men was brilliant, I came away dissatisfied, for the fundamentals I was searching for had somehow eluded them too.

Thinking that perhaps I might find a clue to what I sought in the ancient birthplace of civilization, I journeyed next to India and later to Tibet, and there I found that many searchers had trod the trail that I was following, and that some had actually brought back nuggets of information.

Sometimes these mental "nuggets" were gilded with the

trash of superstition, or wrapped in filmy mystery, but *they worked* in many instances, and if one can press a button and flood a room with light, I do not care how fantastic the theories may be about the origin and composition of electricity.

World War came and on the battlefields I had ample opportunity to apply the knowledge I had gathered and the results were heartening. Then came peace and freedom and the opportunity to settle down to research and experimentation in order to analyze the mass of data collected and to coordinate the results into practical treatment.

I found that there is indeed a key basic regulating factor within the human body—a vital core or soul that is both the originator and the regulator of every mental and physcial function of the body. This factor remains perfect in its knowledge throughout life; it is only when it is interfered with that the organism is prevented from defending itself against bacteria from without, or from repairing damaged or worn out organs or tissues within.

The first of these two factors is the *nutritional* condition. Despite the perfection of the basic core or "soul's" knowledge, the organism must have a supply of all the essential materials it needs from without in order to produce the hundreds of special products it must manufacture within. In Western society there is almost always a long-standing deficiency in several of the more subtle food factors or vitamins as well as other posssible deficiencies.

The second and most important factor is the *emotional* or *mental* factor. All mental responses, if repeated often enough, tend to become automatic, and *every emotional reaction, mild or severe, has immediate repercussions in every tissue, cell and process of the body.* Thus our attitude toward life as a whole determines the *kind* of mental responses we generate toward the innumerable experiences of life, and these responses have an immediate effect on every organic tissue and cell of our bodies. An emotional imbalance becomes automatic and chronic and the perfect core of intelligence within is kept in a state of extraordinary emergency and prevented from carrying out its normal workaday functions of assimilation, elimination, repair and defense.

A great variety of diseases can be produced in the laboratory rat by withholding from its diet one or more vitamins; a great variety of diseases may be produced by keeping the animal in a state of fear or anger; combine both of these factors and the effect is lethal.

Out of these findings evolved the *"A" Reactivation Treatment,* in which this book plays an important part. By a systematic study and application of the principles set forth in the following chapters, that mental substratum sometimes called the subconscious mind can be reeducated into a new, true perspective of life. This in turn will generate the logical harmonious responses so necessary to emotional stability, thus freeing the basic core or soul from its state of extraordinary emergency and enabling it to attend to its normal business of repairing the body.

This is a work that *only you can do* and there are no really effective substitutes for this personal effort. You will not find the task of reeducation easy, for nothing worthwhile is easy, but if you want to return to a condition of perfect health and happiness, and consider that state worth working for, you will have no difficulty in setting aside the small amount of time daily that is needed.

All human life is but a process of gaining new experiences and fitting these experiences through memory into a rational pattern. The process of re-education before referred to must be carried forward in a certain rational order or sequence, from the ground up as it were. The chapters of this book are arranged in that order and you should master each of the chapters in its logical sequence.

The nutritional factor we must leave to a nutritionist in your vicinity or, if you cannot have the services of a specialist in nutrition, your own doctor is the safest substitute. The diagnosis of a nutritional deficiency is extremely difficult and is possible only to one with sound basic training and long experience, and as the long absence from the diet of a certain vitamin sometimes causes a loss of the person's ability to assimilate that vitamin, it must be supplied by concentrating dosages intramuscularly, therefore, advice regarding that phase of the "A" Treatment cannot be given to advantage in book form.

2

Beginnings

"The Gods gave a microbe a drop of water and in it he lived. They gave an ant an half acre of land, and he prospered. They gave a tiger a forest and he formed an empire and became an Emperor; they gave a man the Universe and all the knowledge therein. He entered of his own free will the dungeon of Dogma, shut his mind to truth and slew and starved his brothers."

Book of Right Feeling

Out of the swift succession of great discoveries made within the last score of years in the science of physics, one mighty fact has emerged which confirms triumphantly the ancient inborn faith of all mankind. The Theory of Relativity has proved to us that the material universe with its nebulae, suns and planets, including this earth of rock and soil and water, is but the *observable part*—the condensed dewdrops as it were—of a great rational unseen cosmos of which space and time are but a tiny aspect.

Since the first dawn when humankind stood erect and raised questioning eyes to the heavens, we have instinctively and vaguely half-known that fact, and we have given to the vital, activating principle of the unseen universe many different names: Brahma, Tao, Jehovah, God, and many other names which have been used to describe IT. Inasmuch as all of these names tend to narrow the concept of this stupendous power in the mind of the thinker to that of the concept of a little glorified *human*, we shall use hereafter the term "The Absolute" in speaking of It.

This immeasurable, harmonious, orderly system we call the

universe, then, may be looked upon as a great loom upon which the unseen Absolute weaves Its tapestries of creation. On this earth men and women are the threads It uses, cause and effect the shuttles, and the tapestry unfolds behind us in the pageant of history.

Yet each individual thread is also spun according to an individual pattern, and upon reviewing our past we are often surprised to find that no event, however trivial, has been altogether without significance. Apparently unimportant events have often proved to be causes which produce profound effects upon our own lives and the lives of those near to us. Sometimes it has looked as though we were shouldered and jostled by these events into a pathway that we had no intention of treading; but always in retrospect we can see that the pathway was opened before us for our own greater good.

As you shall learn later in this book, desire, often latent and outside the realm of our consciousness, is the helmsman who steers us through the sea of life toward the port of fulfillment, and it is reasonable to assume that the desire for health has steered you to this book. Believe then in the ability of this desire to lead you to the food that you are after and then make the most of it.

We have read about many wonderful techniques of healing during the recent decades: marvelously delicate surgical operations which can with the cornea of a dead man's eye restore sight to the living; oil of chaulmoogra by which the dreadful effects of leprosy may be modified; barbiturates through which the epileptic may be aided and the attacks lessened. The greatest stories of healing are still those made twenty centuries ago by the young man Jesus who said, "Wash your eyes in the pool of Siloam," and the blind man came seeing, or to the ten lepers, "Go show yourselves unto the priests...and it came to pass that as they went they were cleansed," or when He commanded, "Hold thy peace and come out of him," and the epileptic was cured.

If those stories stood alone we might feel inclined to doubt them—to attribute them to the over-zealous efforts of scribes and followers to inject into the simple and beautiful teachings of Jesus the astonishing and the miraculous. But miracle cures did not cease with the Master's brief ministry.

Down through the ages these astonishing cures have been going on in every land, and many of them have been authenticated by qualified physicians. Perhaps you have read of the occasional healing miracles which take place at the Grotto of Lourdes in France or at Ste. Anne de Beaupre in Canada. The Christian Scientists sometimes produce a cure that can only be described as miraculous; so do the Mormon elders, through prayer and the laying on of hands. In fact such cures are taking place all the time in response to the efforts of many sincere and earnest workers of many faiths in the Western world. Those who have lived in the East can attest to the hundreds of such cures performed yearly by Hindu Yogis and Moslem holy men.

It should be clear to every reasonable person that behind these "miracle cures" performed in every land, and in every age, by believers in widely differing religions and faiths, that there is a common principle which when properly used can produce a cure in almost every case.

This being so, it should be equally obvious that the person in the best position to bring about this sort of a cure is *the sick person him or herself.*

If, instead of depending upon someone else to operate your mental or spiritual powers from without, you master the principles and operate the powers from within the body you wish to have repaired, the results should be not only more thorough but more satisfactory; instead of learning to depend upon another, you have gained mastery over yourself.

In the hundreds of cases within my experience it has worked out in just that way: the person who has helped restore his/her own health by mastering the philosophy within this book is not only more physically fit, but also happier and more successful throughout life.

3

The Reign of Law

"Man can never attain freedom by burning the law books, but he may avoid chaos by studying them so that he may live intelligently within their protection."

Book of Right Feeling

No earthly monarch could rule an empire except through law. To intercede individually in every little problem which developed among even a few hundred people would be quite impossible, so our emperor, if wise, would call together the lawmakers and have a code drawn up by which each would be ensured the right to go peacefully about his or her own affairs without unfair hindrance or interference.

Such a code would have to be based upon equity and justice; it would be absolutely impersonal and if a subject chose to violate it punishment would follow—as a matter of course—as the natural effect of violating the law. Thus our emperor would be relieved of the burden of judging individual cases and enabled to concentrate upon more important affairs of state.

If it would be necessary to rule even a few hundred people in this way, by law instead of by personal intervention and personal judgment, imagine how much more necessary such a system of law would be to rule this vast and complicated universe in which our planet is but a tiny mote of dust.

We call these laws by which the Absolute maintains order and justice in the universe "natural laws," and we use the few of them that we have discovered in planning and carrying out most of the complicated business of civilization. The laws of chemistry, physics, mechanics and so forth are all "natural laws," and their great value to humanity is in the fact that *under similar conditions they always produce the same effects*. Thus if a certain discovery is made in the field of chemistry, the discovery can be recorded and it will work as well for the chemist a thousand years hence as it did for the original discoverer. So the discovering and harnessing of natural law is the most potent single factor in the building of our civilization.

Natural law is not confined to the material aspects of creation. These laws graduate upward like the ascending notes of a scale from the material aspects of the universe to the mental and from the mental to the spiritual; a universe ruled by law in one of its aspects and chaotic and lawless in others is unthinkable.

The tools necessary for each of us to carve out our own salvation are placed in our own hands. These tools are not the making of exaggerated protestations of love and loyalty by way of religious ceremonies, or the fawning upon our knees for "blessings." They are the study of spiritual law and obedience to it. No amount of praying will suspend the laws of gravitation; neither will gravitation cease to exercise its power because you do not understand it. If you jump off a high building you will be dashed to the ground and injured, whether you understand gravitation or not; the natural laws of the mental and spiritual aspects of the universe work in that same way: you will not be punished FOR your so-called sins—*you will be punished BY them.*

However, we are not held eternally within the jurisdiction of a finite set of laws; we may progress to a higher law which transcends the old ones. The following is an illustration.

Suppose that an embryo chick within an egg possessed conscious intelligence. Its universe would be bounded by the egg shell by which it was encompassed; it might first rail against its confinement, not understanding the laws which held it within the shell for its own protection. Next, it might reason darkly about hells of sizzling brimstone outside the shell for chicks who did not accept their lot, and a paradise for chicks who were "good." To a certain extent it would be right, for the premature breaking through the shell, before the body was fully prepared for the new environment, would be disastrous, and when its body was ready, a very pleasant world of immensely broader horizons and new faculties with which to enjoy life would await it.

Eventually the chick would grow until the environment within the shell would be almost intolerable, and would finally say: "This is the end," believing the paroxysms which caused its beak to peck violently at the imprisoning universe to be the throes of death. Then the shell would crack, and it would find a new and broader life...waiting. No longer subject to the old laws which bound it within the egg, it would experience an expanded universe.

Much of the world's unrest at the present time and much of humanity's unhappiness are undoubtedly due to the fact that the spiritual self has outgrown the old laws; the shell is cracked, the horizon immeasurably widened by scientific knowledge and communications. The scientist has become the prophet of modern times and it seems up to each individual to earn the right to individuality by becoming familiar with higher laws through personal study: first study the science of spiritual navigation, then become the captain of your own soul.

In the East, many of the laws rediscovered by the comparatively young Western science were known before our civilization was born. The religious teachers of India taught the story of creation, based upon what is known as the electron theory, many centuries ago. They also taught the theory of evolution long before Europe became civilized and England produced Darwin. In the same way, they have long taught many demonstrable natural laws of the spiritual aspect of the universe; we shall draw freely from these teachings as well as from Western teachings as we proceed.

It may be asked why, as these laws are known in the East, they have not been generally accepted by Western science. The answer is that the West does not consider the methods of the Eastern adepts provable. A Yogi will spend half a lifetime developing spiritual powers; in order to check on the phenomena produced by their use, the scientist would of course have to have a similar development. Western science demands that all theories accepted be provable by the laboratory method, with equipment anyone can operate.

It is in designing this checking equipment that the difficulty lies. Eastern adepts point out that unless a person prove worthy to receive the higher law through self-discipline and study, it is better that s/he be kept in ignorance, for throughout history those few renegades who mastered higher law and used it for selfish personal glorification have brought untold woe to their followers and to mankind as a whole.

Now let us turn to a systematic review of the laws we must make use of in our climb to perfect physical, mental and spiritual health.

4

The Framework of Creation

"The micracle of creation is not that it was done within the framework of natural law, but that it was done at all."

Book of Right Feeling

When the Absolute decided to weave this intricate tapestry called creation, the first step was to devise a loom upon which to weave, a loom that would automatically direct each of the millions of different colored threads into their proper channels for the forming of the grand pattern.

This loom would have to be so sensitive that it would react as perfectly to a floating speck of dust as to the rushing sweep of a flaming sun; it would have to be so responsive that it would supply the needs of life to either a microbe or a human with equal fidelity; all of its myriad responses would have to be automatic and unfailing and cover the whole gamut of physical activity.

Such a loom is what the physicists call "The Field." Simply stated, they have discovered that space itself, instead of being an empty nothingness as it was formerly thought to be, is in reality a mighty ocean of *directive intelligence*, which governs not only the formation of all the physical bodies within it (such as suns, planets, etc.), but regulates their course. The Field also propagates light and all other electro-magnetic phenomena, such as radio waves and more subtle vibrations which are conveyed across billions of miles of interstellar space.

But this is not all; it is *immaterial wave structures* in the Field which organize the neutrons, positrons, electrons and photons into molecules, crystals and solid bodies we call matter, and it is the unseen "pilot waves" within the Field which guide the physical development of all living things into proper channels. There is no recorded case of a honey bee tuning renegade and living according to the life pattern of a wasp; neither can an angleworm follow the life pattern of a caterpillar, or a deer that of a wolf; only the *spiritual intelligence* within humankind is capable of directing the intelligence of the Field, so far as this earth life is concerned, and we have succeeded in doing so only to a slight degree. All other living things are bound to it.

From now on I shall call the Field by the more descriptive name of Universal Mind, for that is what it is in effect. I shall have more to say about it in a later chapter, but before we leave the subject for the present, you should clearly understand that the Universal Mind does not command—it merely responds to the commands of living things.

In New Zealand there is a fjord called Milford Sound where the acoustics produced by the surrounding cliffs and mountains are so peculiar that if a ship's foghorn be sounded or a cannon fired, the sound is caught up, amplified and hurled back and forth for several seconds, producing many weird effects. In the same manner, the Universal Mind catches up the strong desires of living things, amplifies them and answers them by bringing about a grouping of conditions which will enable the desire to be fulfilled. I shall have more to say about Desire later, and I shall point out now that manna will not rain down from the heavens in response to your desire for food. But the war hero and aviation pioneer, Eddie Rickenbacker, who was once adrift on the Pacific on a life raft without food or water for twenty-one days, testified that the desire for food, expressed in prayer, did bring a seagull to light upon his head, thus probably saving the lives of him and his companions, and that it brought a swarm of small fish to loiter around his raft so that they could be caught, and that it brought about a rain squall when they were dying of thirst. Of course, desire must be held steadily until it is fulfilled, for every time it is relaxed or changed the Universal Mind ceases its efforts to respond.

5

Energy

"We live as we control energy—when it controls us we die!"

Book of Right Feeling

Behind every action is energy. In fact, action is but a demonstration of energy. In earlier days physicists thought that phenomena could be divided into energy and matter, but now they know that matter is also energy; the world upon which we live is simply a condensation or denser concentration of the energy existing everywhere in the space around us. So now there is no matter but two states of energy: kinetic energy, or energy in action, and potential energy, or energy at comparative rest. This energy is made up of an infinite number of electrons and protons.

The sun's energy over the ocean causes a certain amount of water to ascend as steam and then as clouds; the energy of the wind carries this water over the land, where cold air currents (more energy) condense it into rain; the rain is pulled back to earth by the earth's gravitational energy.

Consider now the energy manifested by nature. Take a great tree for instance, and suppose that you had to pump by hand for just one day, the moisture and nourishment used by it from the ground to its topmost branches, forcing this sap out through the tiniest twigs, to every leaf and bud of the whole immense structure; you would find that you had expended immense amounts of energy; multiply this daily amount of energy by the number of days of the tree's life—sometimes more than a century—and you will be surprised at the amount of energy which is working silently and unnoticed around you. You can then multiply that energy by a great forest, and to the grasses, the herbs and wild flowers, the insects which pollinize the flowers and the energy of the winds which scatter their seeds, *and you will realize that not only nature but the entire universe is but a gigantic demonstration of energy.*

Science tells us that the sum total of energy in the universe is constant and can neither be created nor destroyed, but can only be changed from one form to another, from potential to kinetic and from kinetic to potential. For instance, a pint of water contains a certain amount of potential energy, which is released or changed to kinetic energy as steam; this steam escapes into the atmosphere and becomes condensed again into potential energy as water. The water which was changed to steam is not lost—every molecule of it returns to its original source.

A ton of iron ore is dug from the ground by the utilization of energy; it is drawn to the smelter by more energy, and the energy of coal is added to smelt it; electrical energy is also added and harnessed to machinery which processes it, and it is made into watch-springs, stoves, gun barrels, wash tubs, automobile parts and dozens of other articles to enable us to more effectively harness other forms of energy. Eventually those articles are discarded as they are worn out and the energy of the atmosphere breaks them down to their original form as rust or iron oxide, which again returns to the earth. Therefore, you may gaze at an automobile or any other complicated machine, and say that this is crude iron ore and other crude natural products which have been changed by energy.

6

Intelligence

"Even an unclean pig knows WHAT *it is doing. but man should know also why he is doing it. Although many care not for reasons."*

Book of Right Feeling

Let us define intelligence as *the purposeful direction of energy.* According to that definition, we see evidence of intelligence everywhere. Minerals display a low form of intelligence and the selective attraction and repulsion displayed by them might easily be compared to the love and hate of humans. Then we see a higher order of intelligence displayed by vegetable life—a still higher order of it displayed by animal life, and the human intelligence may be yet higher.

Now behind all of this purposeful direction of energy we have called intelligence, there must be *something* "directing." An angleworm does not deliberately eat a hole into the ground because it wishes to let oxygen, etc., into the earth that plants may be better nourished. Neither do bees deliberately pollinize wild flowers so that a more abundant crop of vegetation might make life easier for animals who live upon it. Yet, with unvarying energy and industry these lower forms of life go about their labors from birth to death. In the chapter "The Framework of Creation" I have called this directing intelligence "Universal Mind." Let us see if we can give some sort of an illustration: imagine a microphone into which you could speak certain orders, and then imagine the recorder repeating those orders to a horde of employees. Now imagine each employee being equipped with a radio receiving set and your microphone with a sending set. Now eliminate the idea of words altogether and imagine that your employees react wholly to the vibrations transmitted by the microphone. Now it becomes an all-encompassing sea, surrounding this world and filling all space, permeating all material bodies and instructing each according to the part it is to play in the scheme of things; imagine these instructions becoming fixed and automatic through constant repetition in the minds of the species, the cells of vegetation and the crystals of minerals. That may give a crude idea of this directing force.

7

Spirit

"The light of the sun strips night of its secrecy, and makes all things known, yet, that very light protects its own secrets—pry into them and your eyes become useless, persist and it blinds you. Bask in his light as you will, but gaze not too long at the sun, my children."

Book of Right Feeling

To attempt an explanation of the Universal Mind and Absolute Spirit would be presumptuous. An ant would have a better chance of explaining the cerebral processes of Albert Einstein. It is birthless, deathless, not subject to the dimensions of time, and It is the source from whence all knowledge springs. We can know a little of a great artist by the pictures painted; we can learn a little about a scientist by the discoveries made, and a writer betrays a little in his/her books. By knowing that little we can learn to love them. In this universe we see everywhere before us the works of the great Scientist-Artist-Teacher, and by the knowledge so gained we may learn to love. With that we must be satisfied.

There was a time—the post-Darwin period—when atheism became fashionable, because that great scientist was able to prove that humans live in an evolving environment and are a product of evolution. Because we live according to a physical law, there could be no Absolute. Yet today, many great scientists are among our most deeply religious people: not religious in the narrow sense of belief in a personal God, possessing all the human defects of jealousy, anger and hatred, who bestows favor upon the elect because they sing hymns once a week, but religious in the sense that our investigations have led us to a deeper knowledge and a deeper appreciation of the beautiful completeness and indescribable thoroughness *with which every tiny problem of nature has been or is still being worked out.*

I do not think it necessary to prove the obvious. There can be no reasonable doubt in the minds of the intelligent that behind this intricate immense scheme of creation, there must be a *first cause*—an *intelligence* as high above ours as our intelligence is above that of a gnat. So we shall use the first pillar of our "working hypothesis" this proposition: behind all creation, all law, all force, all evolution, and everything which differentiates the universe from a cold, dark, empty and silent void, there broods a mighty first cause, whom we shall call the Absolute.

It is idle to speculate upon the "nature" of this Absolute (First Cause), for to form a clear conception of anything we must establish a standard of comparison, and the Absolute is incomparable. Neither can we consider It in the light of time and space, for time in reality is but an invention of people who called the period of earth's voyage around the sun a "year," then used this period as a mental measuring rod. Time to an insect with a life span of but an hour or so would be entirely different, if he possessed an intellect with which to reason about it. And those reptiles with a life span of more than a century would think still differently about it. In the same way it is difficult for us to conceive of anything being "infinite," for mentally we are chained to the person who long ago marked a stick in twelve equal parts and called it a foot with the graduations inches. We tend to think of all distance and space in terms of feet, miles, meters, etc., but the Indian who measured distance by the day's march had quite a different conception of it, as has the motorist of today, or the astronomer thinking in "light years."

Therefore, when we think of the Absolute as being infinite and "timeless" it staggers our imagination. One may go into a well equipped factory and given the time and patience, may examine the machinery and the manner in which it is organized and then sit down to consider the vast amount of intelligence that went into the building of the machines—even that great amount of intelligent work can be more or less accurately estimated. However, try going for a walk in the country and try to estimate the intelligence manifested everywhere around you in the delicate coloring of the wild flowers, the building of the mighty trees, the busy insects all intent upon the work in hand, the birds cunningly building their nests straw by straw in answer to a wordless desire within themselves, the raincloud shrouding the mountain top, emptying its life-giving moisture drawn from the distant ocean, to make the sparkling stream which waters hills and meadows until it joins its fellows to form the river which again sweeps toward the ocean from which its water originated. Then go out under the stars and observe their order and discipline and consider that the Absolute is the great source of it all. Perhaps after this, the divine attributes to being Infinite, Omnipresent and Omniscient may have a deeper significance, and you may agree with that sage of ancient time who said:

> "A poet, a mathematician or a lunatic might be rash enough to try to explain the great causeless cause; but for me it is enough that I hear him in the singing winds; see him in the sunset, smell him in the fragrance of the moist earth, taste him in the mango's juice and touch him in meditation."

8

The "Descent" of Humanity

"Why, O Rangi, do you deny the possibility of the human mind functioning without a brain after death? Must a butterfly become a caterpillar again before he may think?"

Book of Right Feeling

Humankind fits into this picture in a peculiar and unique way—the body is obviously subject to the natural laws of the universe and is regulated by the great "microphone" (which from now on we shall call "Universal Mind") in common with all other living things. Yet we can, by the magic of our intellect, modify its functioning and make of it our most powerful servant. Luther Burbank, by studying the operation of the Universal Mind, was able to produce entirely new varieties of plants. Cattle breeders have been able to produce new and better breeds of cattle. In fact, practically everything from chickens to babies, from microscopic yeast organisms to gigantic forests of pine have become subject to the directing influence of the human intellect over the operation of Universal Mind or Nature, as we sometimes describe it. Thus we share to a small extent in the creative prerogative of the Absolute, and in this way we may differ from all other living things.

Upon studying the relatively high estate we occupy at the present time, the evidence leads us to the hypothesis that we attained it by the process of evolution. That, like a tiny seed pushing out roots and leaves, stem and branches and a still higher trunk, we evolved from a low semi-animal state by degrees, higher and higher until we attained our present stature.

There is also the equal certainty that "something from nothing never yet was born" and that before evolution, there must have been an involution. Even as a tiny red squalling baby, so helpless that it cannot turn itself over, has within it the potentialities of a great artist or statesperson or scientist, awaiting unfoldment, so had the primitive human within it all the potentialities of our great persons of today and yesterday. In other words, even as the crude chemical elements of the earth, air and water, have within them certain properties which unfold and come into action under the proper stimulus, so our primitive ancestors had within them all the properties and talents of the great people of today, in a dormant state.

All living things in this expanding and evolving universe reach a peak of physical perfection called maturity. Then they reproduce their kind, which in turn repeat the cycle. This is the pattern followed by Universal Mind according to the instructions of the Absolute. And we may be presumptuous enough to suppose that a similar plan was followed in peopling the earth.

Imagine this planet as a whirling mass of incandescent vapors, gradually cooling and taking form under the tireless, ageless action of Universal Mind. Imagine this creaking, groaning, heaving mass gradually being shaped into oceans and continents—the thick clouds of steam and gases which surrounded it like a pall, gradually thinning and sunlight breaking through—first as a murky, yellowish glow, then growing brighter and brighter as the atmosphere cleared, until the new world swung—a silent tiny sphere in the immensity of space.

Now imagine an immense shower of spirit particles falling upon this sphere—the elementary quanta of life—even as the electron is the elementary quantum of matter. Just as every snowflake, under the microscope, has its own individual pattern, so has each of these tiny particles its own inner pattern, a pattern its inner yearning will force it to fulfill as the eons roll by and it learns by successive experiences how best to fulfill that pattern.

In the primordial slime we may imagine this life stuff beginning to work, and from the tiny one-celled monerons we visualize it evolving into mollusks, fish, birds, animals, and vegetable life—all of the immense host working in complete harmony under the dictates of Universal Mind—working even as the tiny coral insect works to build a new island, without knowledge of the Great Plan, but forced to cooperate because of its inner urge for unfoldment. Imagine great fernlike trees and rank vegetation evolving steadily into finer and higher forms; huge, scaly, flying, crawling and leaping reptiles, evolving into more refined and beautiful species until the quality of the young world gives way to beauty and harmony.

9

The Instinctive Mind

"In the voice of instinct thrills a million years of bitter experiences of disembowelings at the drinking place; of bloody tooth and claw battles with the crook-boned men from across the mountains; of sex hunger appeased at the expense of a sleeping rival's life; of the anguish of famine, the suffering of maimed limbs; of all the hatreds, lusts, and fears, in short, born of a savage, primeval environment. Those who have learned to rule their instincts live pleasantly—those whose instincts rule them fill the prisons and the mad houses."

Book of Right Feeling

With a concrete picture in our minds of the great coherent scheme of creation of which we are a part, let us trace the development of the human mind and its bearing upon health and well-being. The first act of the embryo soul, after its descent upon the young world, was to claim from Universal Mind a particle of Its substance—as a plant appropriates a drop of rain drawn up and then precipitated, from the ocean, as its sap. The portion of Universal Mind was charged, like a master contractor, with the task of building a suitable habitation for the soul, which slumbered until the habitat should be completed.

For countless centuries the Contractor labored, a tireless and marvelously resourceful Servant, planning, adapting, altering, and revising the physical body to meet the changing conditions of its environment—devising dozens of little chemical laboratories for the manufacture of substances needed within the body. It planned pigments to act as filters against harmful rays entering the body; elaborated the organs of sight and hearing to catch the vibration known as sound and light. It trained armies of antibodies to repel the invasions of bacteria, to mend broken bones and heal wounds. We shall call a part of Universal Mind the Instinctive Mind, for within it reside, in the form of instincts, all the lessons of the ages.

In the purely animal state, anthropoids were the weakest beast of the forest. They had neither teeth nor claws suitable for combat like the tiger; nor yet the speed of the deer with which to escape the stalking beasts of prey. Those things were not according to the specification of the Absolute. *So the Instinctive Mind specialized a part of itself into an intellect, with the faculty of being able to study cause and effect, to reason from them, and thus outwit the enemy.*

10

The Intellectual Mind

"The snake can hear the earth-song and the lark the music of the spheres, but human intellect hears them both and by science and art interprets them."

Book of Right Feeling

The problems confronting the intellect were at first fairly easy—food, shelter, and security were relatively simple matters and they were partly solved by the Instinctive Mind from its great store of past experiences. Those welled up into the intellectual as desires, fears, prejudices, etc., and guided its every decision; yet in turn the intellect was granted a certain domination over Instinctive Mind. It communicated ideas of danger which at once caused various secretions to be poured into the blood stream to furnish the needed extra amount of chemical energy with which to fight or flee. In fact, the result of its reasoning in every case induced a "mood" and this "mood" furnished the specifications for the Instinctive Mind's chemical balance of the body. It must not be imagined, though, that those two planes of mind are sharply divided. They blend into each other as do the colors of a rainbow, and there is a neutral zone between them which partakes of the quality of each and yet has the functional ability of neither. This is where most of our dreams come from, and it is called by Western psychologists the subconscious mind.

At long last the Master Contractor pronounced the work

complete. By trial and error, through the ages, It had elaborated an individualized spark of Universal Mind which could draw unto itself the necessary chemicals from the air, water and earth, and from them build a wonderful mechanism—the human body—and it was now time for the Royal Master, the Soul, to awake and assume authority.

11

The Spiritual Mind

"But your prison cannot hold me, O little man—on the wings of spirit I can pass between the bars and soar through the powdered gold of dawn to a mountain peak of freedom ye may never reach."

Book of Right Feeling

The Swedish system of gymnastics is planned so that one group of the body's muscles opposes another group to the end that both groups are developed, and in somewhat the same fashion the various phases of human consciousness were unfolded. The Instinctive Mind, by serving up strong and varied desires, caused the Intellect to extend itself to the limit. The development of the Intellect stimulated problem-solving techniques in our prehistoric ancestors. Then as the Intellect matured, it began to ask questions: *"Why am I here?" "What is the purpose of my being?" What happens when I die?"*—and a great, lonely yearning took possession. From that time forward, we could no longer, like other animals, be content to bask in the sun's warmth, fill our stomach with food, gratify sexual hunger and call life good. Intellect told us that life must have a Creator—that this ordered, perfect scheme of things was the work of a great Super-Intelligence.

The effort to establish contact with this Creator, this Super-Intelligence, resulted in the gradual evolution of spiritual consciousness. In humanity as a whole this highest phase of consciousness is still in an embryonic state. Most of us still operate on the plane of the intellectual and instinctive minds. Yet as far back as recorded history and beyond, from time to time have been born individuals who functioned largely on the plane of the Spiritual Mind. Undoubtedly they were evolved as guides and lawgivers to the rest of the race. The unknown authors of the "Upanishads," the oldest and possibly the most profound religious books in existence; Moses and the other prophets of the Old Testament; Buddha; Confucius; Jesus; and Muhammed are a few of the elect. A perusal of any world history will furnish you the names of many hundred more. As man's spiritual mind unfolded, his sense of the reality and omnipresence of the Absolute became stronger and stronger, and a feeling of brotherhood with all living things became very real. Instead of being a lonely stranger wandering in a stark, hostile world, we become a welcome guest in the halls of the Absolute. The carpet of soft grasses Its carpet, the leaping, timid deer, the princely mountain sheep, the trilling birds, the jewelled insects, are, in a sense, the guests of the Absolute as we are.

As a guest it is good manners to treat our fellow guests with kindness and helpful consideration, for by doing so are we not lightening the labors of our Great Host? Neither would it be becoming to take more of the things offered by the Host than a fair share, lest some other guest go without. We must realize also the absurdity of filling our pockets with the Host's silver, for we must soon take our leave and can by no means carry it past the grim doorkeeper.

You, perhaps, have become struck with the fact that the constitution of man, as outlined in this little book, has a striking resemblance to the constitution of the Universe? Our soul is the Universal Mind within us.

12

The Body

"Ye mourn not the plight of a bad ruler, my son, but rather the plight of his people.—Heed then the murmuring of rebellion within your own body and learn to rule it with greater wisdom.'

Book of Right Feeling

We have shown the Universe as a stupendous manifestation of Divine Intelligence, watched over and regulated by Universal Mind within a framework of natural law. We have seen the quanta of life, like a shower of tiny sparks from a sacred flame, descend upon the earth and gradually unfold through the centuries, always progressing, always unfolding more of its innate higher qualities. We have followed our species through this evolutionary process and we have seen the Instinctive Mind learn more about the work of constructing a perfect physical housing for the spirit or soul. We have traced the unfolding of the intellectual mind as the physical housing neared completion. Then at last, we have seen the first glimmerings of the dawn of the spiritual mind as the soul began to awake from its age-long slumber.

Now let us try to form a mental picture of the Instinctive Mind—the Master Contractor, who built the body and who is still on the job to keep it in repair. Imagine a swarm of super bees, presided over by a queen bee of marvelous intelligence and wisdom—imagine the bees, being taught to build their six-sided cells in conformance with a pattern in the mind of the queen—the pattern of a human body. Some are specialized to manufacture a special kind of wax and to build bones with their cells; others are taught to build tissue or flesh, some to build stomach, intestines, lungs, liver, kidneys and other organs: some to make hair, teeth, eyes, ears, etc. As time goes on, these super bees become wonderfully proficient in their work, and as they become more and more specialized they lose the ability to forage for the materials necessary with which to make their wax. So the wise queen solves the problem by specializing still more of the super bees into forming a great network of canals reaching every part of the body where the bees are at work. These canals she fills with fluid, or blood, along which the food-stuffs, gathered and prepared by other specialized groups, may be carried to the cells of every part of the body and the waste products of these cells carried away and thrown off. More bees are organized into a pumping plant to keep this fluid moving; and still more are organized into lungs for the gathering of the necessary oxygen from the air.

One other vital necessity has to be attended to: that is the production and distribution of electrical or nervous energy. So more bees are assigned this work, some forming batteries where the energy can be stored, and others forming the network of nerves called the "sympathetic system" over which it can be distributed. "The sympathetic system" is located principally in the thoracic, abdominal, and pelvic cavities. Through it, Instinctive Mind controls the involuntary processes, such as growth and nutrition, and distributes the "nervous energy" necessary for the carrying out of these functions.

Now our wise queen realizes, in making her plans, that as the Intellectual Mind unfolds, it needs a special network of communication with different groups of bees—one which would enable it to give orders and assist in its processes of reasoning about what is good or bad for the whole swarm of bees. So it devises the cerebrospinal system. It is that part of the nervous system contained in the cranial cavity and the spinal canal—i.e., the brain, spinal cord and the nerves which branch off from the latter. It presides over volition, sensation, etc. Lastly, it knows that it must prepare and specialize a group of super bees as the organs through which Spiritual Mind could manifest itself, to the cerebrum, or brain proper, is developed.

Now, try to form a picture of this immense swarm of super bees, working busily at their special tasks, each one hidden in its own cell, the whole collective effort being directed by the wise queen and coordinated into the physical body. Now each bee's work is necessary to the whole structure, and if a bee, or a group of bees, for any reason stops work, or dies, the whole organization is thrown out of order, and if the situation is not remedied, the wise queen will have no recourse but to order her super bees to leave their cells and together they will fly away in a swarm to an environment they hope will be more favorable—where they will commence again to build a body.

The fact that ordinarily these "super bees" cannot be seen after they leave the body in their swarm at death, does not mean that they are not actual entities; though electrons and protons individually are quite invisible, it is only a question of increasing their numbers in a given space and they can be seen and touched as "solid" matter.

Suppose then that our "super bees" are tiny inframicroscopic units or genes, and that these reside within each of the twenty-five billions of cells which make up the body, forming the governing and organizing power of the cell; such tiny units, though very real, may only amount to one ten-thousandth of the weight of the cells; hence if the whole swarm were abstracted from the body, the resultant loss of weight to the body would be only about a fifth of an ounce. Such a loss of weight at the time of death has been confirmed by Dr. Baraduc in his book, *Mes Morts: Leurs Manifestations.*

13

Prana

"The energy of the swordsmith's flailing hammer added to the energy of the warrior's arm together strike the blow for freedom. A certain sum of energy must be marshalled and directed to accomplish any purpose. O lazy one!"

Book of Right Feeling

Let us now pass to a consideration of that electrical force called "nervous energy," "vital force," etc., by which the cells are enabled to carry on their work and the body to carry on all of its complicated functions. When we wish to raise a finger we put forth an effort of will, if the desire is a conscious one, or if the desire is instinctive, an effort of the instinctive mind, and a small force of this energy is sent to the muscles controlling the finger, causing them to contract, and the finger raises. Every step we make, every word we utter, every tear we shed is a result of this process. So complete is our dependence upon this force, that an eminent Harvard doctor, at a meeting of the American Psychiatric Association, exclaimed: "It looks as if the body is an electrical machine!"

Although the force is electrical, "electrical force" does not describe it any more than "chemical substance" describes flesh, blood and bone. It is a form of electricity generated by the body for its own particular needs. There is not an adequate term in English to express it; therefore, we shall use the ancient Sanskrit term "prana."

As we have said, the body is equipped with an immense network of nerves which form the wires over which the "prana" travels to every cell group in the system, and the life and health of every cell depends upon receiving an adequate supply of it. It is interesting to note in passing that this force is also used by the mind to produce the well-authenticated phenomena of mental telepathy. It supplies the power for the mind's sending and receiving telegraph.

14

Germs

"As the vile thoughts of a drunkard's brain take temporary form in many fearful shapes, so hates and fears and evil passions have generated down through the centuries many other evil shapes—most of which may be destroyed by laughter."

Book of Right Feeling

In a tropical jungle, teeming with life in a myriad of forms, all living things may be divided into two groups: group one are those insects, birds and animals which assist nature by pollinizing flowers, distributing seeds, cropping and pruning herbage, breaking down dead matter, etc.; group two are those insects, birds, and animals which live upon the bodies of group one. Nature uses group two to maintain a balance, for if group one could multiply without restriction, vegetation would disappear, the land would become a desert and all life would find it intolerable. Group two, in turn, have a relatively lower rate of reproduction and are subject to a higher death rate by reason of their ferocious nature.

Now imagine all the animals of this tropical jungle dwindling in size, their bodies becoming simpler and simpler until they are left with the bare mechanism with which to carry out the purpose of their existence—imagine them at last so tiny that one hundred tigers and one hundred deer can stand on the point of a pin. Now imagine the human body as the "jungle" and all of those tiny animals, insects, and birds roaming within its confines.

This analogy is a correct one, for within the human body

are many varieties of helpful germs—from the bacillus which live in the intestines and feed upon the germs of decay to the phagocytes or germs which devour their various kinds of pathogenic bacteria. Normally there is a balance struck between the two groups of bacteria and this is always so in the condition called Health. The body is always bombarded by germs. With every breath we inhale, we extend our hospitality to some of them, with every glass of water we drink we take in others; the paper money we handle, the door knobs we touch, the theatres in which we sit with those suffering from every disease under the sun, and all other phases of our "civilized" life render the propagation of pathogenic bacteria increasingly easy. Were it not for a peculiar defense of the body operated by the Instinctive Mind, it would be impossible to maintain the balance between the groups of the human body, and the race would vanish from the earth.

This defense consists of the ability to produce at will large armies of "antibodies" to combat the invasion and thus maintain the balance of health.

If we put a piece of meat on a plate and leave it subject to moisture and heat and beside that plate of meat place a plate of over-ripe fruit, a remarkable thing will happen: from apparently nowhere a number of blow-flies will appear and settle on the meat and a number of small black flies will appear and settle on the fruit. An ignorant person on viewing the phenomenon might be excused in reasoning thus: "There are blow-flies on the meat—the meat is putrid, therefore the blow-flies must have caused the meat to become putrid" and, "there are small black flies on the fruit—the fruit is decaying, therefore the small black flies must be responsible for the fruit decaying" and thus might conclude that the best way of keeping meat and fruit fresh is to destroy all the blow-flies and small black flies in existence.

It is only when a condition extremely favorable to the breeding of harmful bacteria exists in some cell group of the body, due to trauma (injury) or to an inadequate supply of prana, oxygen or food, or to failure in eliminating waste products, that they can breed with sufficient rapidity to destroy the whole organization of cells known as the body. With a few exceptions, germs should be looked upon as an effect rather than a cause.

To summarize: the human body is an aggregation of an immense number of living individual cells, and each of these cells is the physical house of a life quantum. When these life quanta leave their cells, as bees swarming leave their hives, the phenomenon known as "death" occurs, and the empty cells break up into their original substances.

Each of these life quanta is under the rulership of the Instinctive Mind, which regulates all of the numerous processes and functions of the body to the end that all life quanta may be supplied with the chemical materials necessary to the construction and maintenance of their cells, and the pranic energy necessary for the carrying out of this work. "Harmful" germs are ever present in the body and in one's environment, but are kept in balance by "helpful" germs, and thus is maintained the balance of health.

15

Force and Resistance

"Since 'good' is evil for some and 'evil' is good for some, I hold that these terms are meaningless. The scourge which destroys many enriches the buriers of the dead and creates new opportunities for the living—nay 'good' and 'evil' are but other names for 'force' and 'resistance,' the chisel, and marble of the Gods."

Book of Right Feeling

We have shown that the whole universe is a demonstration of energy, and that what we call matter, is but energy in a different state to that manifested as heat, light, sound, etc. So to give the following picture clearer form, we shall call *force* that energy which acts, and *resistance* that energy which opposes action.

You cannot chip a statue from milk—it has no resistance—you must use marble or some other hard substance which will resist the energy or force of the chisel blows. You cannot give a rope a hard pull unless it is tied to something because it offers no resistance. You cannot give a puff of steam a hard push for the same reason. Likewise you cannot resist unless there is some force to resist. It is this interplay of force and resistance which molds all things and gives to them reality.

The interplay of force and resistance was used to mold this planet from a whirling mass of gases into continents and oceans. Every blade of grass and every mighty tree forces itself upward against the resistance of gravitation. The force of the wind is captured and put to work by opposing the resistance of a windmill to it. In the same way, by opposing resistance to Niagara, its force is made available to turn the wheels of a thousand factories.

When force *is unopposed by* resistance *it loses its creative power, or power to do work, and when* resistance *is unopposed by* force *it becomes meaningless.*

Again, clearly understand, "creation" is neither the product of force nor of resistance. It is a product of both. Force is the "father," resistance the "mother," and the "child" is a higher product, who in his day shall exercise a higher force to a higher resistance (or vice versa) to produce in his turn a still higher offspring. The nations of the world today are grouped into two opposite camps, based upon two opposing political ideals; and out of the conflict, which now rages, will arise a new political ideal. An ideal combining the good points of each with the bad points of neither, and so the work of creation goes on. Now let us see how this principle applies to the individual.

Every one of us has within both creative elements—force and resistance. We grow to the extent that we exercise them; everything we create, whether it be an idea or a new house, is a product of our personal use of force and resistance. If we wish to train as marathon runners we do not conserve our strength and stamina by spending the month preceding the race in bed, but by bringing our vital energy into action against the resistance of our muscles, we improve both to the extent that a superbly functioning body is the result.

In the mental realm we do not prepare for a stiff examination by completely resting our minds; we use their force against the resistance of the subject. The result is a clearer, better mind, and an increased knowledge.

In considering the principle of force and resistance, however, you should realize that a certain balance must be maintained: a hurricane might easily destroy and sweep from its path a flimsy windmill, as Niagara would burst apart a flimsy dam—even as a weaker force might expend itself vainly against a resistance too great for it to move. The "irresistible force meeting the immovable body" would result in the cancellation of each. However, when a weak force expends itself against a superior resistance, it automatically brings about a regeneration of itself as a stronger force. Just so, a futile resistance when overwhelmed has a rebirth as a stronger resistance, and so each grows until a creative balance exists and the new "child" is born.

A weak man might expend his energy against the resistance of a hundred-pound weight day after day, but each time he would discover that his energy had increased a little until eventually he could overcome the resistance of the weight—physical strength would be the result. *A strong character is one whose spiritual and mental force and resistance have developed to a creative balance, by continued exercise against each other, and is never the result of being shielded from temptation.*

16

You

"The universe with its millions of stars—the laughing earth with its teeming life only exist because you are. *Could* you *be utterly destroyed, my son, the universe would resolve itself into black nothingness—so far as you would be concerned, creation would be at an end. So* you *are the most important thing in existence—to yourself."*

Book of Right Feeling

Now my friends, on the wings of imagination we have soared together through the measureless void of inter-stellar space, and then swung backward through the reaches of time to the travail and birth of a new world. We have, by virtue of our wings, floated above and watched the slow, painful climb of man from a moneron wriggling in the steaming slime of a young world to the scientist sitting at ease in the laboratory among the instruments of today. We know that this climb was made possible by the specialized intelligence residing within each of our bodies—the intelligence which through successive hostile environments learned to revise, adapt, and organize until today the soft tissue, brittle bones, and delicately balanced organs of the human body can outwear the toughest steel. We know that all this was accomplished within a framework of natural law, which formed the loom upon which the human fabric was woven. We know that the pattern of each fabric was laid down by the Master Weaver of each body—the Soul.

This journey was undertaken for a twofold purpose. First, to give you a glimpse of the universe, of which you are a part, from the point of view of causes instead of effects. Second, to show you that down through the eons of time YOU have held the reins and guided your own development—physical, mental and spiritual.

In order to restore your body to the equilibrium of health, you must again gather up the trailing reins and with a firm and skillful hand bring the unruly horses of your Instinctive Mind under control and drive them again on their normal route. But first we must help you realize that you *are* the driver. You must come to a realization of the real *you* before you can function properly.

Probably up to this time you have always identified yourself with your body. In fact you have considered your body *you*. You have said: "I am sick,"—"I am hungry,"—"I have a headache,"—I am tired," etc.; according to your physical reactions "you" were fine, or not so good. A juicy T-bone steak, when you were hungry, gave "you" pleasure, and an attack of indigestion made "you" miserable.

And yet the real *you* can never be sick, can never be hungry, can never feel pain, nor aches, for the real *you*—your Soul—is a spiritual entity and inhabits your body even as you inhabit a house. True, if the roof of a house leaks, if it is draughty and the plumbing and sewage systems are out of order, you might be tempted to move if you could not repair them. So it is if the body falls into a bad state of repair; the Soul might do the same thing if the Instinctive Mind did not repair it. Yet the Instinctive Mind can and will repair it in almost every instance when the command comes from the Royal Master—the real *you*—the Soul.

That is the secret behind every "miracle cure" from the beginning of time. The emotion stimulated by prayer temporarily suspends the efforts of Intellectual Mind and the cry for help goes straight to the Soul, and an astounding, vivifying, and harmonizing of every cell and process in the body is the result. I have been privileged to witness many such cures. Such eminent scientists as Dr. Alexis Carrel and Dr.

Alexander Cannon, in their respective works, attest to their truth.

Again realize that up to now you have been living largely on the plane of the Instinctive and Intellectual minds. You have looked upon yourself as the automobile rather than the driver; you have concerned yourself deeply with your physical reactions and intellectually have tried to solve problems which would much better have been left to the Instinctive Mind. Now let us seek out the real master of the situation—the Soul—the real *you*—and resolve that in the future we shall let it deal with every difficult situation.

In every examination there must be two factors—the examiner and the examined—the knower and the known, the seer and the seen. So keeping this fact in mind, go over your body, in imagination, slowly and deliberately, piece by piece, and you will realize as you lay each piece in an imaginary pile, that that is not *you*. Now turn to your mental qualities; your imagination is a useful servant, but it assuredly is not *you;* your memory is an efficient filing system, but still it is not you; your loves, hates, fears, and other psychological states are interesting, but still they are not *you*. Piece by piece and quality by quality, you can lay everything on that pile, but there will still remain the examiner. Now try to imagine this examiner dead—you cannot! Try as you will, you may be able to imagine your body dead, and even your intellect stilled, but you cannot imagine the real *you* extinguished forever.

At this point, again pause and soberly consider how you have been prone to identify yourself hitherto with your moods and passions. "You were sad; you were afraid; you were worried"—it begins to sound a bit absurd, doesn't it? In fact it would be just as sensible to stand out in a storm and to say, "I am wind and rain," or to sit too close to a roaring fire and say, "I am heat." In reality you could be none of those things, you could merely be the somewhat uncomfortable *observer* of them.

And even as you could remove yourself from the wind and

rain by retreating indoors, or escape the discomfort of the too hot fire by drawing back your chair, you can learn to retreat within the encompassing realization of your true self, from the storms of worry and the scorching heat of emotional conflagrations.

A lunatic is a person who identifies with every mood, and so completely do moods dominate that one loses all sense of true personality and consciousness of the I Am. Just remember that the next time a mood threatens to take possession of you, for every time you allow that to happen you are allowing yourself to lapse into temporary insanity. Not much is needed to send the average person into these lapses—hatred, joy, fear or love—any of these emotions may tip the mental beam on the side of temporary insanity, and while in that condition one often says or does things that are bitterly regretted for the rest of one's life.

Learn to pause at the first whisper of a rising emotion and to ask yourself: "To whom is this idea presented?" and then to answer yourself with all the meaning that you can muster: "It is presented to ME the Soul, the very fountain-head of all pure knowledge who uses this mind as a tool, to find ways and means of adapting and applying the knowledge within to the world without. It is presented to ME the Soul, who unmoved, passionless and serene am the *observer* of all storms of false ideas and passions which whirl about ME like autumn leaves, yet which are powerless to influence or affect ME to the slightest degree, *unless I identify myself with them.*"

This frequent shifting back of the personality to the realization of the true self results in what psychologists term an integration of the personality, and the integrated personality is the foundation upon which all true psychic healing and in fact every demonstration of transcendental power rests. Without an integrated personality, you can accomplish little, but once it is accomplished the word "impossible" loses its meaning so far as you are concerned, so no amount of effort is too great a price to pay for its attainment.

This frequent exercise in self-realization will gradually convert the sleeping truth of the fact into a waking, conscious reality, and when this happens you will automatically think, act and judge in accordance with the highest principles of your true self, and by this the dormant psychic power within will be channeled and brought forth.

As you persist in this exercise, the realization of the real *you* will become clearer and clearer and life will take on a new meaning. You will realize that though the body may be destroyed, the *you* lives on and on forever. Fire cannot burn it, nor can water drown it. For all eternity it is destined to pass through new and higher adventures—each experience unfolding a little more of its innate, exalted nature, until the time will come when you will be as a God compared to the state of even the most highly advanced people of today.

This realization may come as a great illuminating flash, or it may come slowly and quietly like the soft flowing of dawn in the eastern sky. But if you persist it will come, and when it does you will realize that you have won the most priceless possession in the world—a realization of the *I am* of the masters.

In a previous chapter you were told that progress is the child of force and resistance; you must realize that the unfolding of a new and higher state of consciousness must be attained by using that same principle, so be not discouraged if the realization of the *I am* does not come immediately. Time after time, return to the exercise outlined above, being satisfied that you are developing a little more force with each effort, and that the resistance is slowly yielding.

17

The Imagination

"Imagination is the mother of hope, and faith is its offspring. It is the light that penetrates the darkness, the searcher who finds the way through the wilderness."

Book of Right Feeling

The imagination is the ruler of the emotions, the creator of desire, the messenger who continually flits with lightning-like rapidity between the domain of thought and the realm of physiological response. It is the connecting factor between all of the three phases of the human mind and it is man's winged ambassador to the Absolute.

The Imagination has all of the forgotten memories of a million years of evolution. It may draw precedents from the lessons of thousands of lifetimes etched indelibly in the vibratory pattern of the Instinctive mind in terms of pleasure or in terms of pain. It is no wonder then that an animal trainer has a most difficult task in impressing a new external pattern of behavior; this is done, of course by impressing the new lessons in terms of pain (the whip) and pleasure (lumps of sugar and other tidbits, petting). However, this newly cultivated path of behavior only acts in relation to the few special situations created for the act or exhibition, and in all other respects the imagination acts in the same way as it did formerly, so the basic characters of the animals are by no means changed to any appreciable extent.

With the human, however, the situation is different. By the Imagination, acting from the Instinctive Mind, the Simple

Consciousness of the animal *knows*, but the knowledge is in the form of a memory, of either pleasure or pain attached to perhaps thousands of similar situations faced through the eons of evolution. The human, by means of Intellect, *knows that he knows*, and is capable of saying: "Yes, I imagine this thing to be true, but that imagination springs from long forgotten memories of pleasure or pain experienced throughout the centuries attached to similar situations, and as pleasure and pain have really nothing to do with truth, I shall examine it with an open mind."

The indirect objective of all education is to transpose the Imagination from the Instinctive Mind to the Intellect, so that human beings, instead of being swayed by the pleasure/pain memories of a primitive past, in making a decision, are able to imagine and to examine the circumstances detachedly from a viewpoint of impersonal truth; to project the imagination thereby illuminating the probable chain of effects arising from the circumstance in question.

In short, the lower animals and our primitive ancestors used their imaginations *backward in time*, to find precedents of "pleasure/pain" as the basis for future action, while the advanced person uses imagination to *search for truth regardless of the pleasure/pain connotations, and looks forward in time to forecast results.*

Thus, we grow according to our ability to control our imagination, and we are impotent according to the degree that our imagination controls us. Much has been written about the "power of will," but where the *imagination* and *will* are in conflict, the *imagination* invariably wins. Thus the most important task in life is to train the imagination into becoming a faithful servant of will instead of an erratic and dangerous master of it.

The lessons which follow in this chapter will enable you to bring your imagination under control, and to use it as a potent and priceless ally in carrying out your plans for attaining health and happiness.

First, you must understand clearly that while the *imagination* is closely associated with memory *it is not mere memory*. It is more than that: *it is the recollection of the significance of a remembered incident in terms of pleasure or pain*. We might think of the brain as a phonograph record, and the senses of sight, hearing, touch, taste and smell, as needles which record the impressions received over these senses on the record. All of these recorded impressions are from *without*, and were they "played back" by memory, without the accompaniment by the imagination of their significance to us, they would come as unemotionally as statements of statistical facts. However, the *imagination* plays along with these facts a sort of musical accompaniment of meaning; we remember some incidents with pleasure, some with pain, others with horror, or pride, or regret.

If we were limited to "playing back" these old recordings of memory with their imaginative accompaniments, we would be as limited in planning the future as are other animals, and if we were unable to change the imaginative accompaniments to old memories, we would be doomed to always act in a similar manner toward new situations which carried any of the characteristics of old situations we had faced—and as something of the old must inevitably be carried into the new—as the future is but the unfolding of the past in a reverse direction—we would act precisely as other animals act, and never grow spiritually.

We cannot only alter the imaginative accompaniment to old memories, but we may by the use of the imagination, make new recordings of the future: RECORDINGS OF EVENTS YET TO HAPPEN.

These recordings of events yet to happen form our pathways to the future, and if the pathways are made deep enough and do not conflict with deeper established pathways, they will guide the "needle" which leads us into time—and our plans will be realized. The events that we have chosen to record, with the accompaniments we have selected in advance, will be played back to us in reality.

FIRST LESSON

The first lesson in controlling the imagination is to free it from its ancient bondage to the Instinctive Mind, and to train it as an efficient servant of the Intellect; we shall teach it, as a first step, to form a series of visual impressions which it will attach to memory without the slightest emotional accompaniment.

It should be understood that the faculty of imagination is closely linked to memory, and that memory is but a loosely interpreted sequence of cause and effect, from a personal angle. Therefore, in making our series of visual pictures, we shall give to each of them an imaginary cause and an imaginary effect. Later you will be asked to recall these pictures in their correct order.

In your mind's eye, form the picture of a high straight wall which bars your view of the landscape, and picture a narrow arched opening in the middle of it. Make this mental picture as clear and as vivid as possible; see the cracks where the blocks of stone are set upon each other, the brown of the sandstone blocks, and the straight line of the wall dwindling in the distance on either side of the narrow opening.

Now visualize a bent, aged man walking with slow and halting steps toward the gateway in the wall; he has long snow-white hair and a white beard that reaches to his knees—he is very feeble.

Now visualize a lusty child of about seven years who runs up to the old man from behind, seizes his hand and pulls him forward toward the gate in the wall. The old man stumbles and hangs back angrily, but the child continues to pull him onward.

Picture them reaching the gateway, then the old man refuses to be pulled farther—he sets his heels in the ground and strives to pull the child back, but the child is even more determined to pull the old man through the gate.

Visualize the old man growing more and more exhausted, and finally allowing himself to be pulled through the gate. You might call this little farce Past, Present and Future, the wall representing the present, the other side of the gate the future, and this side of the wall the past. The aged man is "experience" and the child "adventure," and you might try to remember with the picture that experience, divorced from the spirit of adventure, is weak, timid, futile and unprogressive, while the spirit of adventure, unrestrained by experience, is rash, reckless and childish, but if the two will step through the gate into the future together, great things will result.

SECOND LESSON

A short news item in a current magazine told of how in a Latin American country, where the public transportation system had all but broken down because of a gasoline shortage, the Economic Minister pleaded for a national campaign of "hitch-hiking." Enraged at the lack of public interest in his proposals, he went out on the street to set an example by hitching a ride himself. For two hours he stood on a busy thoroughfare thumbing, with the result that numerous drivers with empty seats looked the other way, while others hurled insults or gibes at him.

Now let us try to create this item into visual and auditory pictures.

First, picture our Economic Minister as a stout, pompous little man, with a fiercely bristling mustache. Second, form a picture of him addressing the assembled members of his country's government, telling them smugly that he has solved the transportation problem. Third, visualize him holding a press conference. Make a mental picture of a dozen reporters scribbling hurriedly in large notebooks, while two cameramen wait to snap pictures. Fourth, visualize the newsboys rushing through the streets waving their papers with black headlines that read: "Minister Urges Hitch-hiking." Fifth, visualize readers of the paper turning to the other items and ignoring the headlines. Sixth, form a mental picture of our little Minister, dancing and waving his arms on the edge of a busy sidewalk, while cars stream past in the street; try to reproduce the *noise* of the traffic with your imagination as well as the visual images, and hear the sound of the derisive horn toots which greet him. Seventh, visualize our Minister walking slowly homeward with bowed head.

Now undoubtedly that experience to the Minister was poignant, because it is mixed with emotion: hatred, contempt, sadness, self-pity and anger probably all played a part in causing him to work out a revengeful repressive measure in place of his campaign of voluntary sharing, but you who visualize the facts utterly divorced from their emotional and personal content, will realize at once that the Minister failed in his campaign because he did not take into account the fact that "*People only change their well established habits when forced to do so by an inner or outer necessity.*" Had he taken this single fact into account, he might have (1) created the "inner necessity" by a carefully built-up campaign of publicity, dramatizing the necessity of sharing rides, or (2) introduced a carefully worked out, just decree such as suspending the licenses of those who, with empty space in their cars, refused to give rides and so imposed the "outer necessity." Thus we might say that his emotion was responsible for a harsh, unjust measure, which called up hatred toward himself in the people, which affected in turn his future in politics, which might have affected the destiny of his country to some extent, and the destiny of his country of course affects, to some degree, the destiny of every country. So it is that uncontrolled imagination is dangerous, and its effects can easily undo a lifetime of careful building.

THIRD LESSON

Take a short news item, any one will do, from your newspaper, and visualize it in the same manner: First, make a vivid mental picture of each of the main characters, then visualize the situations in their proper sequence which make up the news item; carefully keeping your pictures free from any hint of emotion; finally, try to arrive at an understanding of the principle which caused the events in the news item to happen in that way. Remember, the important thing is to practice converting the printed words of your newspaper into a series of vivid mental images—a "mental movie" that you can recall at will. Practice making these little "mental movies" of everything you read; you will be astonished at the ease with which you will be able to recall them, and at the clarity with which you will be able to form definite, unemotional opinions about them.

Begin keeping a journal of the most important events of the day, forming each of them into a little mental movie, and jotting down your opinion as to their significance behind each of them. All of this, of course, is for the purpose of bringing about an interplay between your imagination and your intellect, under the control of your will. It is a long job. However, you will begin to get some results in the form of clearer thinking right from the start.

Many magazines are ideal for practicing the above lesson. You might commence with the first item in, say, *The Reader's Digest*, and make a mental movie of it, then each day take a succeeding article until you have read it through. After you have finished the magazine in this way, commence, without looking at it, to recall each "movie" in all of its details from first to last merely, opening the magazine if you become stalled.

After finishing the magazine in this way the second time, write in your journal the principle you have learned from each item. Write these principles as concisely as the principles of the first two lessons given here are written: "Experience, divorced from the spirit of adventure is weak, timid, futile and unprogressive, while the spirit of adventure, unrestrained by experience is rash, reckless and childish. When the two unite great results may be expected." and "People only change their well established habits when forced to do so by an inner or outer necessity." The main point to watch is emotion. If you find yourself taking sides, either for or against any person or situation in your mental movie, before you have clearly examined every fact bearing upon it intellectually, wipe out your movie and start again. So long as there is a single trace of prejudice or emotion, you may be sure that your Instinctive Mind is tainting your judgment.

After you have practiced this exercise for one full month, and have arrived at the point of progress wherein you can calmly and dispassionately review even the most soul shaking events without emotion and sift them intellectually for the lesson they contain, begin using the method on yourself.

SELF ANALYSIS

Recall detail by detail, without emotion, the most painful experience of your life, but instead of visualizing yourself as the central figure in imagination put an utter stranger in that place. Pass each incident before your mental screen as detached from emotion as though it had actually happened to this utter stranger instead of you, then strive, as in the other exercises, to draw the lesson from it.

Continue with this exercise of self analysis, until every painful experience within your memory has been dispassionately dissected and robbed of its emotional content, and the true lesson of the experience substituted for the old pain accompaniment. If you succeed in doing this, *you will have robbed your instinctive mind of the materials it uses to create your moods of unhappiness*, and you will notice a new elan and lightness of spirit replacing the old moroseness and anxiety of your life.

CREATING DESIRE

The next exercise for your Imagination is to train it in the art of creating desire within yourself. Begin first with imagining some article of food; visualize it until you can see it exactly as it is served; imagine the aroma from it, and intensify this until you almost believe that you can actually smell it. Next imagine the taste of it—make it so real that your mouth actually waters for it—then take another article of food and another, and another, until your whole appetite is stimulated.

I shall have more to say about desire in the next chapter, so I shall close this one with a brief summary of what we have learned about operating the imagination:

First: the imagination can be brought under the full control of your Intellect by practicing the exercises in visualization as outlined.

Second: if you succeed in bringing the Imagination under full control, you have to a very large extent won not only the battle of the past, and freed yourself from those pleasure/pain memories which limit your ability to cope with the present, you have to a great extent *won the battle of the future also*, for outer events are only important to you in the measure that you react to them, and *if you control your imagination, you control your reactions.*

Third: you have a million years of pleasure/pain memories tucked away in your Instinctive Mind. Do not expect to overcome them in a few days or even in a few weeks. Any prize worth winning is worth a long patient effort; you should make the practice of visualizing events unemotionally a part of your daily life—build it into your thinking mechanism until it becomes automatic.

18

Desire

"The thrill of battle, the nobility of self-sacrifice, the joy of anticipation, and the bliss of overcoming are all woven into the cloak of desire—your accomplishments shall be great or little in accordance with your desires. He who ceases to desire, ceases to live."

Book of Right Feeling

Desire is the great moving cause of all action and reaction. The desire for life (self-preservation) and its children—the desire for food, shelter and security—along with the desire to mate, and thus preserve the race, are fundamental urges. But had our desires stopped there, we would still be living in caves and shivering in ignorance. Desire, like our shadows, ever recedes before us and we attain one only to have another and higher desire form within us to take its place. As the desires of our Instinctive Mind are gratified and a measure of security is afforded us, the desires of our unfolding intellect begin to assert themselves. We begin to crave intellectual food. We use reason on the problems surrounding us, invent new apparatus, pride ourselves upon our scientific achievements and devise new systems of government. We strive for an intellectual Utopia, where humanity shall be as efficient at the business of living as a hive of ants or bees.

Then dawns spiritual desire and we begin to realize the importance of the individual—to understand that any system of government which enables the individual to earn the necessities of life in peace and to cultivate the higher self according to inner desires is a good system, and we bend our efforts toward eliminating the obstacles in the way of this individual development. It is well to note in passing that the conflicting desires of individuals, or groups of individuals and of nations, form the force and resistance which give birth to new and higher desires.

Now let us return to a consideration of how all this affects *you* and your problem of attaining a state of health.

You have been shown in a previous chapter that your Instinctive Mind has within it the accumulated experiences of the ages. It is the most resourceful, most adaptable thing in existence. It resides within you, repairing all damage to the physical structure in a marvelously resourceful way. The reason it has not cured you in this particular instance, if you are sick, is not that the disease germs are stronger than it, but for some reason it has lost the *desire*, or urge to stamp out the germs and repair the damaged tissue.

This loss of desire, it is well to point out here, might have been caused at least in part, by your Instinctive Mind being denied the right materials, in the form of food, from the outside to do the repair work, for to work its miracles of biochemistry by which antibodies, hormones, enzymes and all of the other specialized organic compounds are produced within the body to maintain it in a state of health, it must receive from the outside the raw materials of proteins, carbohydrates, fats, minerals and vitamins. If the materials it receives by way of food are unsuitable, the Instinctive Mind is forced to either laboriously adapt them as "makeshifts," or to rob less vital parts of the body of those already in use, and to use them in the most vital areas.

In primitive man, when the Instinctive Mind needed a certain combination of materials for the repair of the body, it gave its owner an insatiable craving for the food which contained them, which would send him hunting everywhere until he found that particular food. However, with the development of the Intellect and its product, our highly specialized, industrialized and artificial system of "civilization," this natural appetite atrophied and may no longer be depended upon—no longer are flavors guides, they are camouflage.

However, assuming that such vitamin, protein and mineral deficiencies as might have existed have been made up on the advice of your doctor, or a good nutritionist, the next step is to awaken the desire for health again in your Instinctive Mind.

The practice of medicine consists of the introduction into the body of various drugs and substances, which, experience has proved, cause the Instinctive Mind to throw off its lethargy, to produce the antibodies to fight the germs, to throw off toxic products, and to repair damage. But your doctor, skillful and scientific though he be, is trying to stir your Instinctive Mind from without—while *you* are within—*you* inhabit the body, and without *your* wholehearted cooperation his efforts will be of no avail. I repeat, your first act of cooperation must be to awaken the *desire* within your Instinctive Mind for a healthy body. When this desire becomes powerful enough, never fear, results will follow as day follows night.

A story is told of a great Hindu healer who had exhausted every resource of his art on the son of a Maharaja, who suffered from paralysis, without avail. One morning he carried his patient out to an ornamental pool in the palace gardens and without ceremony dumped the youth into the water and held him immersed despite his frantic struggles. When his patient was about to lose consciousness, he lifted him from the pool and placed him upon the grass. When the youth had recovered sufficiently to sit up, the healer asked:

"My son, while I held you beneath the water, what did you desire?"

"Why, O Guru," the youth answered wonderingly, "I desired nothing but air."

"Well, when you can desire health with the same intensity that you desired air when you were beneath the water, you shall be healed."

The story illustrates the principle I am trying to explain. You are where you are and what you are according to your desires—or lack of them. Much is made of reason, but if you are absolutely honest with yourself, you will find that you act as you *desire* to act and use reason to justify yourself for so acting.

You say, "But I desire to get well." But do you really *desire* to get well? Or do you merely wish you were well? Do you desire health as the youth in the story desired air, or have you more or less adjusted yourself to invalidism and passed the responsibility for your recovery on the shoulders of your doctor? No, I am not scolding you for your lack of desire. The human Instinctive Mind, as has been said before, is the most adaptable thing in existence, and often, when the intellect is deeply distressed by the realization that we are suffering from supposedly "incurable" disease, it saves our sanity by withdrawing the *desire* for health and by adapting itself to limiting as much as possible the further inroads of the disease. Then, we say, the disease is "chronic."

It is this writer's experience that such an adaptation is much more difficult to overcome than the actual disease itself, for the healing powers of the body are only brought into action with full intensity when the situation, either through physical pain or mental craving for a restoration to health, becomes intolerable.

The Instinctive Mind has no reason to lift the body it serves out of an adaptation that is comfortable—an adaptation wherein the patient has adjusted himself to being waited upon, and wherein he derives a keen pleasure from sympathy. Why should it wish to alter this pleasant situation and for that matter, whey should a doctor wish to alter it?

There is a dramatic appeal about illness and we are all prone to fall under its spell, particularly when our lives are not in immediate and obvious danger. We see ourselves as martyrs, bearing up nobly under suffering, and we absorb (what we are pleased to pretend is) the homage of our friends as a sponge soaks up water.

We browse aimlessly from doctor to doctor in the hope that

we may find one who will cure us without calling upon us to sacrifice any of the pleasant little habits of repose and irresponsibility that we have accustomed ourselves to, and certainly without calling upon us to work mentally from within, and to cultivate new and realistic attitudes toward life.

It is not unusual for such "adapted" patients to violently deny that they are feeling better even when mechanical and chemical checks prove that they are vastly improved, and this is not to be wondered at, for they realize that recovery means losing the role of hero that they have played in the little drama, and a return to the prosaic routine of life. Thus, almost without realizing it, *many patients will fight to retain their disease.* This is what doctors call malingering and it forms one of the greatest problems in the treatment of chronic disease.

The cure depends upon helping the patient to realize that the real heroes are those relatives and friends who continue to sacrifice and smother their own natural instincts in order to serve. The average person has not much patience with sickness naturally; we feel that we belong to life and we resent having to adjust ourselves to the slow death of a friend or relative. That we are noble enough to deny emphatically that this is so, by no means alters the fact, and our subconscious resentment often sours and warps our whole life when the condition is extended over a period of years.

The above should not be considered in the light of a blanket denunciation of the chronically ill—it is merely a truthful statement of a little understood psychological condition that is a natural by-product of chronic disease—*a condition that only the patient can successfully treat,* and almost every intelligent patient is quick to admit its presence, when it is pointed out, and once having realized the presence of this enemy, is ready to defeat it BY CREATING WITHIN THE SELF THE DESIRE FOR HEALTH.

Now let us pass to a consideration of how the desire may be re-established:

First, return to the exercise in the last chapter. Meditate

upon the reality of the *you* until you can look upon your body and its Instinctive and Intellectual Mind as *your* vehicle or instrument. Now visualize the life quanta each in its separate cell, working like the swarm of bees mentioned in a previous chapter, each tiny intelligent spark moving and acting in response to the *desires* of the Instinctive Mind.

Now form as vivid a picture as you can of your body in a state of perfect health. Imagine the exhilaration of perfectly responding muscles in long swinging walks through the crisp autumn air. Imagine camping trips in the mountains—of awakening after long nights of perfect sleep, washing in a cold mountain stream and sniffing with delight and combined fragrance of pine, wood smoke, coffee and bacon cooking over the campfire. Imagine the delight in climbing to the high places and of looking out over the blue valleys, and sungilded mountains, and of the feeling, "This is my world. I belong to it and it to me."

Imagine the pleasure of being able to perform your chosen work with zest and enjoyment—of finishing each day's labor with the thought, "I am doing my best and fulfilling the purpose of my existence." Imagine the reawakening of keen interest in the affairs of your community, your country, and the world—of viewing the great, moving, shifting drama of life with joy, not forgetting the comedy of it also.

Realize again the absurdity of identifying yourself with your pain and your discomfort—that it would be just as sensible to identify yourself with the squeaking hinges of your back door. Determine that from now on, instead of thinking of yourself as an individual in pain, you will strive to realize the real *you*, the Spiritual entity who is beyond pain. The *you* who created the nervous system over which the pain travels to the brain you created, and resolve to re-establish the harmony of your body through your creative power—that you are the master of pain, not its servant.

As has been mentioned before, realizing the above once in a while will not work a miracle, for down through the ages your Instinctive Mind has woven a little fairy story around the "material overcoat" called the physical body—a fairy story such as a child weaves around her beloved doll—endowing it with a personality and a reality that it by no means possesses, and this has been repeated over and over with millions of variations until the very cells of your brain have arranged themselves into a belief pattern, forming a pattern of thought which is played back automatically every time your illness is thought of, making you think that you are ill instead of your body.

You must work constantly to destroy the old belief pattern by superimposing over it the *Truth*, that the real you can never be sick—that it contains within itself all of the wisdom and knowledge by which it built the body and manifested the mind in the first place, and having this knowledge, is amply able to repair the structure it has built, if your mind is made to assist instead of thwart it.

The "new record" of truth should play back something like this whenever your illness is forced upon your attention: I—the real ME—am perfect. I built this elaborate overcoat called a body, together with its brain and nervous system, and I designed the hundreds of special laboratories within, which manufacture all of the hundreds of special chemical compounds needed to restore and to maintain this body in a state of health. Having built this body in the first place, I know that I can repair it, for my resources are unlimited—I shall remove all interference from this work of repair by controlling my imagination and through it my mind and my emotional state. I shall keep every mental process harmonious and tranquil, happy and relaxed. I shall turn this illness into the greatest benefit that has ever come to me by making it a Royal Road to SELF-REALIZATION.

Practice this until it all seems very real and possible, then, speaking as a master to a servant, say to your Instinctive Mind: "You again desire to bring the body to a state of perfect health. You desire it as a drowning person desires air. You shall awaken every cell of this body into new activity to rid itself of toxic products, charge itself with energy, and to set about the labor of repair with renewed vigor."

Do not be dismayed if an immediate "miracle" does not take place. Remember, every cell in your body has adapted itself to a state of chronic invalidism. It has attained a new equilibrium and, in order to go ahead with the elimination of the disease, this equilibrium must be upset—the balance destroyed—and every cell must cooperate in attaining again the equilibrium of perfect health.

Remember also the principle of action and reaction. Every action brings about an exactly equal reaction. If you persist in these exercises, if you use the force of the real *you* and your imagination against the resistance of the balance, or equilibrium of disease, you can feel absolutely certain that each time you exercise you are developing a little more force. If you continue, the equilibrium of disease must certainly be destroyed, and that of health restored.

19

Silence

"The prattle of the market place is but the inane monkey chatter of the jungle, which serves only to distract the mind and exhaust its divine potentialities on trivial issue. Find ye the silence, O searcher for wisdom, for in the silence wisdom speaks."

Book of Right Feeling

According to the psychoanalytic point of view, a strong desire rejected by the conscious mind will persist in the unconscious (or, for our purposes, the Instinctive Mind) until satisfied. Frequently, the desire has not been consciously satisfied because it is forbidden by our own social or ethical codes. Our Intellect is in conflict with our Instinctive Mind. The Instinctive Mind (Freud's "unconscious"), failing to achieve its goal through conscious means, does so unconsciously, often with astonishing physical repercussions; e.g., the repressed need for sympathetic understanding may bring about physical symptoms which would naturally elicit sympathy, thus fulfilling the unconscious need of the Instinctive Mind without the attendant guilt that would accompany a conscious plea for such understanding.

Psychoanalytic treatment of repressed desire consists of a long examination by the patient and analyst of the patient's past to uncover the cause of the early repressed need(s) thus, according to Freud, freeing or curing the patient. The Unconscious (Instinctive Mind) then no longer harbors that desire, and the patient can allow him/herself to become well again.

I am explaining the above because it illustrates, in reverse, so to speak our next lesson—the lesson of silence.

If you faithfully practice the exercises in the two preceding chapters, desire will be created in your Instinctive Mind for health, and that desire will continue to motivate it to use all of its marvelous resources to fulfill that desire—*until you talk about it, when it will automatically be released from further effort.*

Contemplate the previous paragraph; it is applicable to many aspects of life. Although anticipation often constitutes half the joy of an endeavor, excess verbalization *before the fact* can often vitiate the anticipated pleasure. Silence can indeed be golden. The brilliant talker is seldom the brilliant doer.

In a later chapter you will be told at length how the suggestion of others may affect your recovery, but here it should be understood that there is a peculiar quirk in the Instinctive Mind of most people which reasons something like this:

"To admit that I do not understand a thing, is a confession of mental inferiority; therefore, I shall refuse to believe what I do not understand and I shall try to destroy it in the minds of those near to me, thus restoring my position as an equal by bringing their knowledge down to the level of my own."

The above is not deliberately reasoned out, of course; usually it is purely instinctive. Generally there is at least one member of each family who poses, usually without justification, as the "superior" member, and he or she strives to maintain that position at all cost, usually by the above method. The suggestions of this individual given in the form of covert sneers, jibes and cheap "wise-cracks" is a far deadlier poison to the chronically ill person, who is fighting to readjust mentally to health, than any toxin produced by disease.

Then again we have the "Calamity Janes" who revel with morbid gusto in the agonies of others; these people instinctively regard the illness of others as special melodramas written for their entertainment. Add to the list those who exhibit a morbid fascination with death and dying and you emerge with an unfortunate number of people who contribute to the disease (dis-ease) process. I have been told by those who should know, that an invitation to attend an execution is seldom turned down, and the popularity of those news items which contain lurid accounts of lynchings and other spectacular murders is evidence enough that the majority of people derive enjoyment from the contemplation of agony—so long as it is happening to the other fellow. Of course the "Calamity Janes" would be chagrined, and even horrified, to be told that they possessed this type of mind, and they would most vehemently deny it, for their conscious minds usually succeed in finding a "high moral reason" for yielding to the sadistic urges of the Instinctive Mind.

Nevertheless, this type of person is a real menace to the success of a patient's fight to overthrow adaptation to chronic invalidism, and the less such a person knows about your plans and hopes, the better chance you have of finding fulfillment.

You will learn in the chapter on "Psychic Influence" that even the silent incredulity of another may have a certain amount of power to slow your progress, so learn to *deliberately conspire with yourself* to fan the flame of health within the silent forge of your body, and to strive for the realization of your true self—*by yourself,* for that is the only way possible.

The foregoing should not be interpreted to mean that you must not be frank and outspoken with your doctor and his assistants, however, for their basic understanding of your problem depends upon your full confidence and frankness.

First, then, you must find the real you; second, that real *you,* by dictating to your Instinctive Mind, and by calling to its aid all the resources of your imagination, must implant within it the *desire* for health. Third, you must keep absolutely silent about your desire for health, and that which you have planned for, until it is attained.

20

The Emotions

"You talk of the age of reason—bah! Life is not reasonable, my Son—love, hate, rapture, sorrow, ecstacy, despair, serenity and misery are emotional states, and they are the language of life. Your musty tomes on logic are but fit for the unburied dead."

Book of Right Feeling

Somewhere within the warp and woof of your physical and mental makeup, every experience of your life that has aroused emotion is recorded.

Every experience you undergo now, however trivial, which arouses emotion, either contributes to or retards your recovery.

What you are today is largely due to the countless emotional experiences of your life, yet those experiences which aroused no emotion within you have left no imprint. They have been discarded and forgotten, like seeds which failed to sprout.

In fact, the history of civilization is written in the ink of emotion, and world-shaking political events are in reality but emotional upheavals.

Events, to the individual, are either great or unimportant in proportion to the emotion they arouse; and all of us are heroes, scoundrels, saints, sinners, philanthropists, or misers in accordance with the type of emotions we cultivate and entertain.

Use your imagination as vividly as possible on each of the following experiences as described; you will find that different types of emotion are stimulated. Imagine:

. . . the blushing dawn lifting the mantle of night from a

sleepy bestirring world with its twittering birds and subdued half-noises of awakening life . . . the serenity of a desert night, when we renew our souls by communing with eternity, made manifest in the blue liquid vastness of a bejewelled sky . . . the laugh of a little child . . . the gentle smile of a loved one . . . soft music which carries us on enchanted wings to the magic land of tranquility . . . the echoing salvoes of a thunderstorm in the mountains and the hissing slash of rain . . . the thunder of drums, the eerie savage wail of the pipes, and the measured tramp of marching feet . . . pinch-faced, starving children with great eyes and wonderful faces . . . a shattered pile of debris that had once been the home of love and laughter, and beside it the mangled body of the babe still clutching a broken doll with its tiny, bony hand . . . the mourning of an autumn wind among the branches and the dying leaves fluttering back to their mother earth . . . ineffable sadness of a sunset . . . the poignant music of Beethoven . . .

You will see that the above experiences arouse within you very different types of emotion: tranquility, desire for action, pity, horror and sadness.

Let us return now to the proposition that every experience you undergo, however trivial, which arouses emotion, either contributes to your recovery, or retards it.

All of the fear emotions, such as anger, hatred, worry, jealousy, malice, horror, etc., cause the adrenal glands to pour forth an excess of adrenalin into the blood stream, the pancreas to pour forth an excess of glycogen, the heart to pump faster to distribute these products to every part of the body, the lungs to work faster to gather more oxygen, the digestive organs to cease or slow down their functioning, the blood to leave the capillaries and to concentrate in the main vessels or vice versa—all the repair work going on within the body is suspended. The reason for this is that the emotions were always a prelude to action. When fear was generated, it was an order to the Instinctive Mind to prepare the body to fight or flee and it answered by the process outlined above, withdrawing energy from all functions not immediately essential and concentrating in the motor system of muscles. When action followed, as it usually did, the excess amount of energy expended followed by a slight period of fatigue forced the body to relax and the normal chemical equilibrium was restored. Today it is different—action seldom follows those emotions and the excess chemical products are eliminated slowly, thus keeping the whole physical mechanism choked and toxic, slowing down repair work within the body to a snail's pace, or stopping it entirely and leaving the mind in a "jumpy" chaotic state.

The lower emotions, at the present stage of human evolution, have outlasted their usefulness and now they exist only as dangerous, irrational, roving factors in the Instinctive Mind. The higher emotions tranquilize and harmonize every cell in the body. The Instinctive Mind is released from a state of extraordinary emergency and is allowed to concentrate its efforts upon peaceful affairs—repair work and digestion, assimilation and elimination, glandular balance, the destruction of harmful germs, etc., take up its time.

"Yes," I hear you say, "That is easy to understand; but how *am* I to control emotions which come unbidden, like hurricanes, and sweep reason and self-control before them like straws in the wind?"

The answer is this: "You can, if you will, learn to control your emotions as easily as you control the motor of your automobile by turning the switch on or off. That switch in your Intellectual Mind is called the imagination. *Unless it is turned on, no emotion can be aroused.* You can examine any experience intellectually, pro and con, but unless you turn your imagination on, it is powerless to affect you either for better or for worse, and you can mentally toss it aside and forget it as easily as you would toss aside a worthless article which had accidentally come into your possession.

Doctors, judges, and others whose daily experiences are particularly harassing, instinctively learn to control the emotions through switching off of the imagination. Soldiers must learn it or perish. Tens of thousands of hypochondriacs, neurotics, paranoiacs, and others who exist in a living hell among us could cure themselves and live happy, wholesome, and useful lives if they learned to operate the switch of imagination at will.

The trick of operating the switch is easily learned: first, form the habit of daily correcting your perspective by finding the real *you* as outlined in a previous chapter; then, remain on guard intellectually throughout each day, examining coldly every experience which comes to you, and asking yourself, "Is this experience constructive and would I enjoy having it written in an illuminated scroll and hung in my parlor—or is it destructive and better discarded?"

If it belongs to the first category, then switch on the imagination full force—enjoy the experience to the uttermost. If it belongs to the second, then toss it aside and switch your imagination onto something pleasant. After a time, your imagination will become trained to seek the constructive and the beautiful—to listen to the golden symphony of spirit and to close its ears to the discords of fear, hate and disintegration, and life will begin anew.

In a previous chapter I emphasized that it is not only dangerous to identify yourself with your moods, but that your moods are the psychic thunderbolts which generate the emotional bolts of lightning which may destroy the entire moral structure built up over many years by self-discipline.

However, occasionally even the most self-controlled person becomes swept into the wild current of an emotion which seems to come literally "out of the blue," and it is again necessary here to point out that when this happens, instead of accusing yourself for your "weakness" and going into a dive of despondency, you should strive to benefit by the experience by practicing detachment—by calming yourself as much as possible and asking, "To whom is the emotion presented?" then answering, "It is presented to ME the observer, who stands solid and unshaken like a lighthouse amidst the flying spume and howling winds. The emotion is no more to ME than the gale is to the lighthouse, and I am no more identified with it than is the lighthouse with the storm—it is not ME and I am not it."

21

Suggestion

"You talk of independence—freedom—liberty—and there is strong magic in your words, yet what have YOU originated independently of the thoughts, suggestions and ideas of others, obtained from your associations, study or education. Only the Gods may claim originality and independence!

Book of Right Feeling

A group of four young doctors were holding an animated discussion as they walked toward their university in a sleepy, old English town. The subject of their discussion was the power of suggestion. One held that the only virtue in any medicine lay in its indirect yet powerful suggestion to the patient, conveyed by the temporary masking of symptoms or cessation of pain; another argued that the reaction of the drugs themselves brought about the improvement and that suggestion had little or nothing to do with it.

"For instance," this youth continued, pointing to a group of laborers engaged in digging a trench in the pavement, "It would be quite impossible either to harm or to help any one of these men by pure suggestion, as it would be impossible to influence any normal, matter-of-fact individual."

To settle the argument they decided upon a plan: first, one of their number approached a blond young giant who was swinging a pick, and after gazing at him earnestly for a minute or two said, "I am Doctor Jones, and I could not help but notice your dangerous state of health. Don't you think you had better go home to bed and send for your doctor?"

"Why," the astonished laborer asked, "I have never had a sick day in my life. I am feeling fine."

"Very well," answered the "doctor" ominously, "you are certainly going to be sick—and very sick." Shaking his head, he left.

Half an hour later, another one of the students, apparently hurrying past the spot where the laborers were working, glanced casually at the young giant, then with surprise and horror stamped upon his face, stopped dead and addressed him. "I am Doctor Smith, and I would like you to know how a man in your obvious condition can continue working. Surely, you feel a violent pain here?" touching him on the temple, and here?" touching him on the stomach.

In due course the third student came by and repeated the suggestion, but by that time the young laborer was actually suffering from a violent headache and nausea. Before the fourth could drive the suggestion still deeper, the victim of the experiment had gone home and to bed. The results might have proven serious had not the four gone in a body to his bedside and explained.

A little thought given to the subject will convince you how our lives are largely ruled by suggestion. Every book, every magazine, and every newspaper we read is built upon the principle of suggestion. An author writes a book or a story around a theme chosen for its value as suggestion. The suggestion may be, as in the novels of Charles Dickens, that certain social injustices should be corrected, or that it may be the inherent right of youth to independence, or that crime does not pay—but always, the suggestion is there, and so it is with much that comes over the media. Each newspaper has what is called an "editorial policy," and that policy may be that either the Democratic or the Republican Party are political morons, or that a certain foreign country has deadly designs upon us and so forth; the editorial policy of each paper is different and reflects, usually, the ideas held by the editor. Thus, a man running for governor of a state on a platform of pensions for the infirm, aged, and destitute, may be, according to the editorial policy of one paper, a great-hearted humanitarian—the pioneer of a new and higher order of social justice, and according to the editorial policy of another periodical, a charlatan of the worst type, who makes political capital out of the misery of the aged and penniless, a crackpot visionary who would bring about chaos with his wild ideas. Each paper is reporting the same news, the same speeches and addresses, but each gives it the twist to suit or fit its editorial policy. As we said before, tens of thousands of readers accept the suggestion conveyed by this twist without question and repeat it as their own. Advertising, with which we are bombarded—in magazines, newspapers, and in other media—influence by suggestion our buying habits and our standards of living. Not only are we, in our daily lives, swimming like fish through a sea of suggestion and allowing our lives to be governed by it, but our whole personalities are largely the result of suggestions we have accepted from the days of our childhood onward.

As a child, you were told that if you did "so and so," you were a "bad boy or girl," but that if you acted in a different way, you were "good." So it became a matter of importance to suppress those impulses which caused you to act in a manner called "bad," and to give rein to those which were followed by "good" acts. As this little book has not been written with the problem of child training in mind, we shall not go into the folly of suppressing emotional impulses in children, once those impulses have been developed; we will assume that our education was but the adopting without question of an infinite number of suggestions.

We can take four educated, rational people, each belonging to a different religion, and the chances are that although each will concede the rationality of the others on all other matters, each is privately convinced that they are all very much misguided on the subject of religion—except himself. *The reason for this, of course, is that each has accepted, without question, a series of suggestions on religion, and that the suggestion to each of the four upon that subject was different.*

Enough has been said, perhaps, to convince you of the immensely important role that suggestion plays in your life and the life of your community, your nation, and the world. So now let us return to the individual and study the mechanism of suggestion, so that we might learn to control it, instead of allowing ourselves to be controlled by it.

In preceding chapters, it was shown how the Instinctive Mind specialized a part of itself into the Intellectual Mind, whose function is to exercise the faculty of reason and to solve the problems existing outside the body. It was also explained that the Intellect secured the cooperation of the Instinctive Mind by inducing moods which served as patterns for the chemical balance of the body.

Now the Instinctive Mind cannot reason—cannot argue. It accepts as facts such suggestions as are dropped into it by the Intellectual Mind, and proceeds at once to put them into execution.

In the illustration we gave at the beginning of this chapter, the Intellectual Mind of the young laborer reasoned thus:

"Three doctors, all of them disinterested, have said that I am ill—therefore, I must be ill, even though I don't know it. One of them said I must have a pain in my head and my stomach. It will certainly appear before long."

In answer, his Instinctive Mind said:

"For some reason, it is necessary to make this body ill—to produce a pain in the head and in the stomach—to hear is to obey," and the suggestion was put into effect.

By this illustration, you can easily see how important it is for the Intellectual Mind to exercise discrimination in passing upon the ideas presented to it, both from within and without. Most thoughts presented to our Intellectual Minds have two parts; one part is to inform us, and the other part is to influence us into doing something about the information. The part of the idea which informs us is neutral but the other part—the affective part—is usually the expression of someone else's pleasure/pain emotional reaction to the first part, and it seeks to arouse the same emotional "pleasure/pain" reaction within us.

Suppose that the idea of an old country house is presented to us. One person presents the idea as "a draughty, tumble-down old barn." Without examination of the property in question, or discarding the affective part of the idea by intellectual discrimination, we would store away in our Instinctive Minds a very unfavorable impression of the place indeed. If, on the other hand, it had been described to us as "a quaint and charming old place," the impression stored away in memory would be quite different.

Again, suppose that someone presents the idea of Mr. Jones, as "that lazy, gossiping old rascal." Our impression before we became acquainted with that person would be quite unfavorable, whereas if he had been first described to us as "an easy-going friendly old philosopher, interested in his fellows," our impression would be quite different.

Thus, our mental attitude toward almost everything is conditioned by the emotional response *of others* toward those things; our relations to very many things and experiences are determined by the affective responses we have been gullible enough to absorb, long before we have actually come in contact with the things or experiences ourselves.

If we would be freed from mob thought (or lack of thought) and mass hysteria, we must form the habit of discarding all the affective parts of the ideas which are presented to us, we must use our intellects to absolutely amputate the emotional fraction from them, and to store them as statements of fact, to be later classified according to our own experience with them. In the examples quoted, we should have stored merely the idea "an old country house located in a certain place" and "an elderly man whose name is Jones."

Most of the negative affects are passed to you with the idea of arousing your hatred or fear of a thing before you have the opportunity to find the truth about it for yourself, and most positive affects are passed to you with the idea of causing you to form a favorable impression without examination. So beware of them both. *Remember, no idea in itself is either good or bad, except its emotional affective content makes it so;* you cannot afford to become an instrument to be played upon at will by either the hysterical rabble-rousers or the doleful players of dirges—*you owe yourself independence.*

Now, of course, suggestion is just as potent, and even more so, when used positively. The Instinctive Mind is inherently constructive and its main work is to build and sustain—not destroy. Therefore, positive suggestions of strength, harmony and well-being are of immense value in assisting the body back to the state of health.

In fact, the possibilities of suggestion as a therapeutic agent have hardly been explored in the Western world, although in India, Tibet, and other Eastern countries, whose high marks of civilization were reached as ours was born, suggestion is used not only as a tremendously effective healing agent, but to produce many astonishing phenomena such as clairvoyance, telepathy, levitation, telekenesis, etc. In fact, the powers of the Instinctive Mind, when properly invoked, seem almost unlimited.

Suggestion, to become effective, must get past the Intellectual Mind into the Instinctive; so long as the Intellectual Mind harbors any doubt it is automatically prevented from being passed along to the Instinctive Mind to be acted upon. In passing, it is interesting to note how the Great Healer stressed the importance of this point.

"What things soever ye desire, when ye pray *believe* that ye shall receive them—and ye shall have them."—"All things are possible to him who believes."—"If ye have *faith* even as a grain of mustard seed . . ." and so on. In fact, this principle is so stressed throughout the entire New Testament that it is surprising that its significance has not been more fully understood and practiced. For in clear and unmistakable language it offers the key to health and self-mastery—the key to the Kingdom of Heaven—which is within.

Now there are three methods of applying suggestion: direct suggestion to another, which is only effective when the patient or subject has implicit faith in the truth of the suggestion;

auto-suggestion, the effectiveness of which depends upon a clear realization of the real *you*, upon the absolute intellectual acceptance of the suggestion, and upon the control of the imagination and through it, the emotions; hypnotic suggestion, which has been employed in the East from the dawn of time. In the nineteenth century it enjoyed a brief vogue in Europe. Anton Mesmer, an Austrian, began practicing it and the results obtained were so startling that they created a profound stir in scientific circles. Mesmer's methods were taken up and improved upon by such great doctors and scientists as Lebeault, Charcot, Freud, Milton and Erickson. Dr. Braid of England performed over one thousand surgical operations, using hypnotic sleep instead of anaesthetics, giving his patients suggestions that their healing would be rapid and clean. To this day, his results have never been equalled, in spite of the improvement in asepsis and surgical technique. On his deathbed Freud expressed his regret that he had failed to pursue hypnosis as a means of curing his patients.

The very spectacular success of this "new" method of treating disease, however, proved to be its undoing so far as Europe was concerned. Quacks and charlatans were quick to take advantage of the then "mysterious" influence. Amateur "healers" sprang up like weeds everywhere, and for a time threatened to eclipse organized scientific medicine. "Magnetizing circles" became the fad of society—vaudeville and variety shows staged acts in which people were hypnotized and made to perform absurdly. All this ballyhoo and furor caused the medical community to withdraw from it in disgust and in self-defense, lest they be classed with the "amateur healers" and vaudeville showmen. The medical associations were forced to protect the dignity of their members in the same manner, and frowned upon any doctor who used hypnotic suggestion in his practice. And so, believe it or not, for a century, humanity was deprived of a method of healing which would have meant happiness to millions, only because of the astonishing success of that method.

There is nothing weird or mysteriously occult about hypnotic suggestion. The operator merely lulls the subject's intellectual mind into quiescence by monotonously repeated suggestions: "you are growing sleepy, your eyes are growing heavy!" etc. When the intellectual mind has ceased to function, he talks straight to the subject's Instinctive Mind (the unconscious) and suggests that it speed up the work of repair in the body. Every mother hypnotizes her babe as she croons it to sleep with a rhythmic rocking; and every psychologist who has sat in church and listened to an intoned sermon realizes that the originator of this method of preaching, in the misty past, knew well the methods of hypnosis and employed them with great effectiveness.

Rhythm of any sort is a powerful hypnotizing agent. Music, for instance, is used by the Whirling Dervishes and when a state of ecstasy is attained, they thrust knives through their cheeks and muscles, withdrawing them without spilling a drop of blood, and the wounds close immediately. All dancing, in fact, whether it be the war haka of the Maori or the graceful tango of the Argentineans, is a method of inducing hypnosis. In this lies its popularity. The dancer is lifted temporarily above the sordid world of struggle and worry to a magic land of romance, glamor and happiness—who shall say that this life is not made richer and more beautiful by the memories of that land, even when the dance is over? And who shall say which of these lands is more real?

In closing this chapter, I wish first to speak a word of warning to those who might be tempted to experiment with treatment by hypnotic suggestion:

First, remember that every suggestion dropped into your Instinctive Mind will be acted upon. Therefore a doctor skilled in physiology and pathology, who understands what must be done by the Instinctive Mind to overcome the disease, is a safe practitioner. Never submit yourself to the experiments of amateurs!

Second, always have a person, whom you can trust, present

during the treatment as a witness, so that later you can, if you wish, learn exactly what was suggested. And understand always, you cannot be hypnotized against your will. Hypnosis can succeed only with the full cooperation of the subject.

Now let us return for a moment to auto-suggestion. In order to successfully employ this method, you must first, through the desire to get well, find the real *you* as previously explained; *you* must direct the imagination upon the condition of health you wish to attain—the imagination used thusly will arouse the emotion which will set the Instinctive Mind into action to make the suggestion a fact.

Great care must be exercised to avoid negative suggestion—for instance, do not say to your Instinctive Mind:

"I shall immediately set about repairing the lesion on my lung!" Instead, you should say:

"My lung is perfect. The whole of my wisdom and resources are employed upon it—every cell in it is filled with life and prana and I will stimulate them to the limit of creative activity. I am perfect and each quanta of life is within its cell ready to spring to work at my command."

Do not say: "I shall at once proceed to cure my toothache." Rather, say: "I shall cause all the nerves of my face to function perfectly and harmoniously."

The reason for this is if you suggest curing a disease, it is suggesting to your Instinctive Mind that the disease actually exists, so it will continue the existence of the disease and at the same time carry out as best it can the suggestion regarding fighting the disease. So the net result would be an increased virulence of the disease plus an increased resistance to it. In other words, *all* suggestion must come from a positive point of view to be effective.

22

The Removal of Limitation

"You cannot see the fly in your eye, O my brother, as you can the gnat on your foot; thus you cannot see the true significance of the affairs you are engaged in, unless you look backward over the events which preceded them."

Book of Right Feeling

Many times in your life, perhaps, you have marvelled at some great athletic feat of strength, endurance, or agility; perhaps you have even wondered why some humans should be endowed with this ability to far transcend the efforts of their fellows. In the same way, you may have often wondered, perhaps a little enviously, at the extraordinary talent displayed by some members of your own trade or profession, who seem to be able to handle the most difficult problems with ease.

Yet, experimental hypnotism proves conclusively that the qualities mentioned above are not possessed merely by a few "supermen" of the race; they are the common property of each and every one of us. Almost any good subject, under the influence of hypnotic suggestion, can be made to develop apparently superhuman strength. I have seen a small, slender man in this condition given the suggestion that his body was rigid, and that nothing would bend it; the back of his head was placed on one chair and his heels on another, while a two-hundred-pound man sat on his middle without bending it in the slightest degree. I have seen a young woman, a music teacher, given the suggestion that she was a tight-rope walker, and she moved across the rope with perfect grace and balance, yet in her waking state she could not move forward two feet on the rope without falling.

In the same manner, almost any artist or artisan, under the influence of hypnotic suggestion, can produce work far transcending anything he can normally produce. Now what is the explanation of this phenomenon?

In previous chapters, we have traced the development of the Instinctive Mind down through the eons—we have seen how it passed through life after life, learning a little in each period of physical existence and adapting its body in accordance with the lesson learned. Desire was the inner urge which drove it onward, and pain was the guide which kept it on the right path. If a child touches a hot stove, the shock of pain he experiences will prevent him from repeating the experiment, and forever after, he will have a healthy respect —or fear—of a hot stove. In this same way, the Instinctive Mind learned its lessons of preservation, for it loathes pain.

With the development of the Intellectual Mind, however, and the birth of self-consciousness, a new element of pain entered into the picture—mental pain; as the mental pursuits engaged in became more important to the race than the former, purely physical, activities, this element played an increasingly important part in our development. We became exposed to humiliation, ridicule, disappointment and frustration, and each of those things brought mental pain. So the Instinctive Mind acted in precisely the same manner toward avoiding a recurrence of mental pain as it did in avoiding a recurrence of physical pain: it developed a "fear" reflex against repetition of the incident which caused the mental pain. Furthermore, if, despite this instinctive fear, we persist in trying to repeat the incident, the Instinctive Mind endeavors to fill our time so full of trivialities that we will have neither the time nor the energy to accomplish our main objective: those aptitudes which, in hypnotic experiments, enabled us to excel in certain pursuits are withheld.

In sickness, this principle operates in the same manner.

Almost every chronic case holds a history of great hopes turned to bitter disappointment. Here is a typical case history:

A high school girl of seventeen, who had suffered from a hacking cough for a long time, had a pulmonary hemorrhage in the class room. She was taken home and a physician sent for—the verdict was pulmonary tuberculosis. The shock to this young, intelligent, ambitious girl was severe; all her golden dreams for the future were wiped out, and she found herself staring into the naked desert of chronic invalidism. Her whole soul rebelled and she commenced a frantic search for a cure. Advice from well-meaning friends poured in, and she finally went to the sanatorium.

The first few weeks there were a living hell. To all her plans for "when she was cured," she was met by pitying smiles from the other patients, or by headshakes. At last, in desperation, she spoke to the doctor, who told her that it was unwise to dream of "cures" insofar as tuberculosis was concerned. It was much better to think in terms of "arresting the disease."

She left the sanatorium under what was practically a sentence of death, and for a month tried, by a life of hectic gaiety, to forget, only to break down completely and be taken to another sanatorium.

Then followed twelve years of bitter disappointment and frustration. She tried everything that was suggested to her, only to be disappointed. Then she turned to Christian Science, but by that time, her Instinctive Mind shrank from the task, wishing to avoid the mental pain of more disappointment. At last, with one lung completely gone, and half of the other badly affected, she fell into a state of cynical resignation and waited patiently for the end.

Now, a little thought over this true case history will show you, who have read the previous chapters of this book, where the real trouble lay.

The Instinctive Mind has charge of all work of repair in the body, and performs this work with unbelievable resourcefulness if allowed to do so without interruption. You know also that the emotion of fear causes a very definite interruption or suspension of all repair work for the time being. Now, the Instinctive Mind reacts the same, whether the fear is of mental anguish or physical pain; so here we have the paradox of a patient being defeated in her battle for health by the very earnestness of her efforts. It was as though a child who had once touched a hot stove, tried to force his hand back to it time after time, each time trying to tell himself that it would not burn him again; each time he experienced the pain of the burn, his Instinctive Mind reflex against the act would be that much more powerful.

Again, remember that when the conscious mind and the Instinctive Mind are at variance, *the Instinctive Mind usually wins.*

In the example previously given of the young woman walking the tight-rope under the influence of hypnotic suggestion, she was able to do so because, with her conscious critical faculties temporarily suspended, and her imagination under the control of the hypnotist, that imagination was told positively that she would walk the tight-rope with ease—her Instinctive Mind released the necessary aptitudes to enable her to do so.

In her waking state, the conscious critical faculties of her mind came into play, and told her imagination: "You cannot do that—for you it is impossible. You would be certain to fall, and not only make yourself ridiculous, but perhaps injure yourself badly." This thought of mental and physical pain promptly swung the Instinctive Mind into action, and withdrew her sense of balance and the other aptitudes necessary to accomplish the feat.

So also with our tubercular student. Each time she tried some new "cure," her critical faculties said: "Well, I suppose this will be only another disappointment." Wishing to avoid the mental pain of such, her Instinctive Mind, whose cooperation in every and any cure is absolutely essential, withheld its efforts and forced the patient away from the course which promised the pain. In other words, her Instinctive Mind had to choose between the possibility of acute mental pain cause by disappointment, and the comparatively lesser pain of chronic tuberculosis, and it chose the latter. It is worth mentioning in passing, that when this young woman was induced to take up the work of reorganizing her mental life in accordance with the principles outlined in this book, she became her own best doctor and an almost immediate and pronounced improvement in her condition took place—an improvement which progressed steadily as long as she was under this writer's observation. During that time she graduated from the condition of a hopeless, helpless invalid, so weak that she could not speak a dozen words without utter exhaustion, to that of a vital, vivacious person who walked a couple of miles each day for exercise, and could discourse interestingly and animatedly upon any subject for as long as it pleased her. In all probability her cure was completed eventually.

It has been explained in a previous lesson that the *Imagination* is the push-button by which the intricate machinery of the Instinctive Mind is set in motion, and again that when the conscious mind and the Instinctive Mind are at variance, the Instinctive Mind invariably wins. The remedy, then, for cases such as those quoted in this lesson, is obvious:

Realize first that the imagination can be a most convincing liar. A soldier awaiting the order to "go over the top" can die a hundred deaths in imagination—and yet come through unscathed. Yet by each of these imaginary deaths, he becomes a little more inefficient until, when the time comes to advance, he is but a blundering, dazed automaton—a walking corpse, incapable of the quick, clear thinking and swift, decisive action which would make his chances of coming through unharmed, infinitely greater. He has two alternatives to that of dying those imaginary deaths—one is to detach his imagination from the situation by forcing it to focus on some other situation or thing (I knew one officer, decorated several times for bravery, who at such times would try to mentally solve problems in arithmetic, and another who would compose jingles); the other alternative is to force the imagination into a clear realization that, regardless of what happens to the body, the real "you" is deathless and cannot be injured.

To return to the problem of using the imagination to overcome the instinctive fear caused by previous failures, one must return in imagination to some occasion in which one enjoyed the fruits of success; we must try to re-create the thrill of joy, the expansion of spirit, the glow of self-approval all over again, then we *must carry that mood to the problem in hand*. If we can do this, the entire machinery of the Instinctive Mind is set in motion to aid us, forgotten or unsuspected aptitudes appear, and success is easy.

To apply this principle to sickness—practically every chronic invalid has had "ups and downs" and must recapture the rapture and exhilaration of that first period of apparent recovery. If we cannot retain that feeling of rapture and exhilaration indefinitely, then we should detach our imagination by turning it onto something else, preferably joyful, stimulating reading, good music, or onto some constructive work, such as writing to friends, etc. We should never allow ourselves to talk of the condition, nor even to think of ourselves except in terms of recovery and health.

23

Psychic Influence

"The crashing monsoon is less terrible than the stealthy plague, because we can see the first and guard against it, while the latter strikes us when we are off guard. Thus the blustering bully may be more lightly offended then he who smiles and remains silent to your jibes—the first may strike with his fist, but that is less to be feared than he who strikes with a thought."

Book of Right Feeling

Were you ever plagued by a certain tune running endlessly through your mind, and then had someone in the same room commence humming or whistling it? Did you ever, apparently without any particular reason, begin thinking of a certain friend or relative, and then go to the post office and receive a letter from that person? Have you ever thought how strange it is that many important scientific discoveries seem to be made simultaneously by two or more people, working without knowledge of each other, in different countries?

I suppose that almost every writer has had the experience of working over a story, only to pick up a copy of a current magazine and find that same theme and plot. So common is this phenomenon that it is behind many of the charges of plagiarism which are bandied back and forth in the literary profession.

During the past forty years, these phenomena have been receiving an increasing amount of attention from Western psychologists. European investigators of eminence, such as Richet and Flammarion, have attributed them to mental telepathy. Dr. Rhine, at Duke University, in North Carolina, has been a little more cautious and tabbed the cause as "extra-sensory perception." But perhaps the most valuable of all investigations was conducted many years ago in this country by the Zenith Radio Corporation. They solicited from their listeners true experiences of telepathic communication, then authenticated each experience by a thorough investigation and, if proven, broadcast it to their listeners. In those experiences were premonitions of events which later actually happened, and knowledge of happenings to friends or loved ones at a distance.

I spent my earlier years among the Maoris of the South Seas where telepathic communication plays almost as important a part in the affairs of the people as the telegraph does in the United States, and later had the privilege of studying the ancient wisdom of Mother India in that country. All this ponderous investigation of a subject proven many centuries ago seems a bit absurd, but if it brings home to the average person the important part that it plays in life and happiness, it will not be without value.

As you read these words, perhaps one hundred or more radio and television programs are passing directly through your body and brain, yet you know nothing about them. But suppose that somewhere in your head you had a little receiving set that you could operate by a mere effort of the will! By a slight effort, you could turn to any program you wished, and could select any type of music from the latest hard rock to the sublime harmonies of Strauss or Chopin.

You not only have such a "receiving set," but a combined broadcast and receiving set in your brain, only it receives and sends not radio or television programs, but human thoughts. This receiving and sending organ is called the pineal gland, which is proven to be something of a puzzle to Western physiologists. They realize that somehow it plays an enormous part in mental development, but most of them think that it is the remains of an "ancestral eye." However, the delicate eye-like mechanism bears out the findings of the Eastern investigators; for such a mechanism to catch and send the electrical impulses of thought would necessarily have some similarity to the eye—the organ for receiving the electro-magnetic phenomenon known as light.

In a previous chapter, I spoke of prana—the electrical force generated and stored by the body, to be sent out over the network of nerves to the millions of cells as the energy necessary for them to carry out their work. I mentioned also that this prana was the energy which operated the body's "sending and receiving sets." Now this is how radio station Y-O-U works.

When you concentrate upon any thought or idea you are broadcasting that idea automatically. Your thoughts are hurtled out into the ether and may be received by others thinking over the same idea. This broadcasting does one other thing—it attunes your receiving apparatus to the same type of thought, and the moment you relax they (the thoughts of others on the subject) commence to flow into your mind. It is interesting to note in this connection that a paper was read before the recent meeting of the American Psychiatric Association, proving that while a person is thinking, he is emitting positive electricity from his brain, and that while he is relaxed or sleeping, the current becomes negative.

Now you can realize by the foregoing the immense importance your thinking has upon your health. First, if you worry, you are "tuning in" on all the worry thoughts in existence and, as was shown in a previous chapter, fear thoughts (worry is slow fear, you know) are taken as suggestions by the Instinctive Mind and it responds by filling your system with unusable products to prepare you for combat or flight. This makes you toxic and suspends healing and digestion; if this persists, it brings on functional disorders which become chronic. In addition, you are squandering your pranic energy which would be much better employed by the cells in their work of healing.

You would not, I know, turn your radio on and tune it in to a horrible discordant chatter—then why tune your mental radio to the same thing?

Practically all of us belong to one of three types of thinkers. First, the worrying type, who remains attuned to fears and suspicions and the woes of the universe, and in consequence lives in a hell of his own creation, and is avoided and disliked by his fellows. Second, the indifferent, casual type who lacks the power of concentration necessary to either send or receive well. Members of this type usually drop into a rut of physical routine and are satisfied to live the life of contented cattle—neither accomplishing much nor failing badly at anything. Third, the cheerful, positive type, who tries to leave the world a little better than he found it, who recognizes that everything is born of force and resistance, and that to accomplish anything one must attune oneself to courageous optimism and like a happy warrior, fight for the sake of the battle.

Before reading this book, you might have belonged to any one of those types—heredity and environment play an important role in the adjustment of your sending and receiving set. It is difficult for a child to attune itself to thoughts of happiness, love and harmony, when it is bombarded by the oral negative suggestions of carping parents. The continual mental telepathic suggestions of fear, discouragement and misery have a profound effect. Again, a child raised in a wealthy household, where there is no incentive to do anything except seek amusement and entertainment, will probably belong to the second category.

And so it goes: the mental adjustment of every person to life depends—or did depend—very largely upon forces beyond his control or understanding.

But that is no longer so for you—you can, by following the instructions given, attune yourself either to the howling of the jackals or the music of the spheres. You will find that environment is an effect and not a cause, and that you can make it what you will, in accordance with your desires. You will find that health is our natural state, and that if you do your part in removing interference from your poor, harassed Instinctive Mind, it and all the forces of the universe will cooperate in re-establishing health and then in maintaining it.

24

Knowledge Harnessed

"What you learn is of interest of your teacher and what you know is of interest to yourself, but the knowledge you put to use is of interest to all mankind. Knowledge unharnessed is but idle vanity and an affront to the Gods."

Book of Right Feeling

I suggest that you read through this chapter briefly at first, then continue to the end of the book, for in this way you will not lose the continuity or "thread" of the exposition. Afterwards you should return to this chapter daily for it is the "key" designed to help you harness the knowledge contained in all other chapters.

There is a difference between merely reading and studying—we usually read for entertainment, but we study to acquire knowledge—yet even the knowledge gained by study is of no practical value unless we learn to apply it, therefore we shall try to combine the study of our subject with simple directions for application.

Of course you realize that the mental attitude you bring to any course of study largely determines the value you receive from it. Knowledge is only appropriated by you when it strikes a responsive chord within, so in our outline of a study method I shall lay down no time schedule, but suggest that each lesson group of chapters be read and reread, until the intellectual acceptance of the first reading is replaced by an inner certainty of *knowing*; when this takes place it is time to turn to the next lesson group.

As the *first lesson,* we shall take up *the attainment of silence.* To be silent sounds like a very simple feat, and in the ordinary physical sense it is, but to attain silence in a metaphysical sense means much more than merely refraining from making a noise; it means the absolute stilling of all thought processes while in the waking state and this is at once the most difficult and the most vitally important lesson of all. The most difficult because the habits of a lifetime must be overcome temporarily with each period of silence; the most important because all other exercises commence with a period of silence and end with a similar period.

First, let us take a brief glance at the mechanism with which we are about to work. The brain is an organ elaborated by Instinctive Mind for the purpose of receiving sensations, sorting those sensations into meaningful patterns, and then directing the activities of the rest of the body accordingly.

The activities of the organs of sense are automatic and continual while in the waking state; our eyes are ranging constantly, our ears and nostrils are alert, and any unusual activity which comes within their ken is at once sent to the brain, where it is catalogued according to the importance of its relation to the welfare of the body.

Now this constant gathering of sense impressions played an important part in the preservation of the race in early days, and at times it is still necessary, as for instance when crossing a busy street. While the brain is engaged in making patterns of these sensations, it is incapable of receiving the deeper and more subtle spiritual impressions from within, or of acting efficiently upon the vibrations of Universal Mind. This is why Nature stills the mind to sense impressions periodically by the method of sleep, and that is one of the reasons why sleep is so essential to human welfare.

The first step in attaining silence is to make the body as comfortable as possible; it does not matter much whether you are stretched prone on a bed, seated in a comfortable chair or upon the ground outdoors—the important thing is that there be no unusual strain in any muscle group.

Second, you should select an object to rest your gaze upon; this should neither be too close nor too distant, else a certain amount of eye strain will result. Now place your fingers upon your pulse and try for a minute or two to catch its rhythm; after getting the measured tempo of its beat, release the pulse and slowly inhale, completely filling the lungs in eight pulse beats; retain the breath for four pulse beats, then exhale, completely emptying the lungs, in eight pulse beats; hold the lungs empty for four pulse beats, then repeat as before until the lungs have been filled and emptied eight times. (An exception must be made to this exercise, however, in patients suffering from pulmonary disease; they should not indulge in any breathing exercises unless closely supervised by their doctor.)

Having completed the breathing rhythm, now deliberately dissociate your mind from sense impressions. You will find this very difficult at first; slight movements and slight sounds will cause the eyes and ears to telegraph impressions to the brain until it may seem as if all nature and every person within earshot is in a conspiracy to prevent you from stilling your mind.

Now turn the mind inward upon self and hold it as still as a placid lake. Here again you will meet with difficulties for little thought-ripples will keep intruding—half formed ideas, scraps of conversation and various other tag ends of mental activity will clamor, or whisper seductively for attention; you must persist in silencing each of these as it happens until, when you become expert, time will stand still and you will have only a sense of personal identity. All else will be a formless, soundless void.

At first you may only be able to hold this silence for a few seconds. Later, with practice, you will be able to hold it for a minute or two, but eventually if you persist in the exercise, you will be able to enter the silence at will and remain in it for as long as you wish.

Even from the first you will come out of the silence with a new sense of tranquillity and power, and this will remain with you for some time afterward. Later you will learn to go to the silence whenever beset by doubts, fears or worries of any description, and you will find that always you will emerge from it with a new sense of serene assurance. The best time to undertake these exercises is early morning and at night, before bedtime.

Second lesson: enter the silence, as outlined in lesson one, for ten minutes, then read over, in the manner explained in "The Evolution of Consciousness," the following chapters: (1) "Spirit," (2) "Intelligence," (3) "Energy," (4) "The Framework of Creation," and (5) The Reign of Law." This is in reverse order to the way you first read them, but I suggest this order purposely.

Now by an effort of the imagination, try to create a series of pictures of the immense system these chapters portray: the Absolute radiating through all space a power greater than that radiated by ten thousand suns, yet so perfect and thorough that even the tiniest insect lives and moves and has its few moments of life by virtue of it. This power is beautiful, harmonious and tranquil, more powerful than a mighty pair of pincers which could burst a world asunder, yet gentler than the soft brush of a moth's wing against your cheek. It is this power coursing through the artist which finds expression in his pictures; through the composer and the musician it expresses itself in harmony of sound; through each one of us it finds expressions in some degree, for all creative work is an *expression of it,* and every manifestation of beauty from the rendition of a great overture by your favorite symphony orchestra to the coloring of a butterfly's wing is an *expression of it.* The more of this power we learn to express, the greater we become, and the less of it we express the lower down the scale we descend towards futility and nonentity.

Mediate upon the all-pervading sea of Universal Mind—the great ocean of constantly operating wisdom that is yours to call upon at will, able and ready to solve your every physical problem. Then visualize force at work, and realize that all this energy in existence is cooperative, helpful and ready to assist you if you but understand the laws by which it operates as you would understand the laws by which electricity works if you were an electrician.

Realize that the same ether of space through which the Absolute transmits creative vibrations from the cosmic ray to sunlight, is *also* in and between every cell of your body and brain, ready *also* to convey the vibrations of your thoughts.

Finally, meditate upon the fact that the laws of the Absolute always work the same way. They are quite automatic and invariable and were put into force so that we may, by using them, attain our fullest expression of perfection.

Third lesson: enter the silence for eleven minutes, then study the chapters: "The Descent of Man," "Instinctive Mind," "Intellectual Mind," and "Spiritual Mind." Now lay aside your book and in imagination try as before, to create the drama of creation. A drama in seven acts.

First, imagine the edict of law flaming out from the Absolute like an immense play of the aurora borealis, this phenomenon gradually fading as it organizes into silent unseen natural law.

Next, in the embrace of natural law, the electrons and protons swarm together to form the nucleus of our planet; then, this nucleus starts on its whirling course, drawing to itself more and more electronic power until it assumes the form of an incandescent mass of whirling, billowing gases, then gradually, cools to a sphere which is rent continuously by great explosions of gases from within. Whole continents heave up from the steaming, boiling sea, taking form for a moment like dream pictures in the mist, then break up into new forms and slide from sight, as the travail of successive convulsions shakes the young world.

Then a bare, lifeless world, bald of vegetation and without any form of life, steams in a steady deluge of rain; obscured by clinging mists and silent save for the shuddering rumble of an earthquake, or the crash of landslide; then the quanta of life descend like a snowstorm of infinitely small pin points of light.

Now the harsh boldness of our world is cloaked with green as vegetation sprouts, each tiny quantum of life striving busily to express itself and design ever newer and better adaptations to its environment.

Last comes the human animal. At first a great hairy creature, but little different in outer appearance to other animals, yet possessing as a life quantum a tiny fragment of the Absolute's substance, thus possessing potentialities of grandeur and power, even as a tiny unlovely seed possesses the potentialities of a mighty tree.

Finally, the recognition by humanity of its selfhood as an individual. The graduation from being satisfied with a life of eating, sleeping, fighting and roving, to one of planning, thinking and sharing with the Absolute the prerogative of creating to a limited extent our own environment.

Think over what has been said of the Instinctive Mind. Many of its instincts are now handicaps, instead of assets; fear, hate and selfishness are no longer of any use to us and should be replaced by courage, fellowship and unselfishness.

Realize that the Intellectual Mind now has new material and a new perspective to base its reasoning upon. Instead of dropping negative thoughts into the Instinctive for action that is no longer needed, and upsetting the chemical balance of the body of peace, power and harmony by so doing, we can adapt to a new order of things, where jungle methods are useless.

Now dwell upon the Soul, and its emanation, Spiritual Mind; all along you have been looking upon yourself as a material body possessing a mind, but now you should begin looking upon yourself as a spiritual being, temporarily appropriating to yourself a constant stream of mineral substances, eaten as food, drunk as water or breathed as air, to provide yourself with a "material body." *You*, this real and spiritual *you*, is deathless. Your existence is more permanent than the ancient mountains or the distant stars. You have within yourself the pattern of a superbeing as far above your present self as a sublime flower is *above a little seed*.

Your inner desires have urged you to your present stage of evolution; vague yearnings and feelings of discontent with things as they were led you onward. You have often misinterpreted these yearnings and tried to satisfy them by seeking material or sensual diversions, and at times they have made you unhappy. Now, you must realize what they are, and by directing them properly use them to bring to flower the perfect picture within.

Any diversion from health, harmony and abundance is in no manner due to this perfect picture within your soul; it is because in some manner the picture has been obscured. Brush the dust from it and keep it constantly before you, then see how eager your instinctive mind is to bring it into reality.

Fourth lesson: this time enter the silence for twelve minutes, then study the following chapters: "The Body," "Prana," and "Germs."

Refresh your understanding of the fact that you are a *spiritual* being and that you own the various phases of your mind, in the same way that a carpenter owns his tools. Before your spiritual mind commenced to unfold, the operation of your Instinctive Mind was largely automatic, and your Intellectual Mind was more or less a mere extension of your Instinctive, but with the dawning of soul consciousness, those phases of mind look more and more to you, their master, for orders and become increasingly reluctant to act upon their own initiative.

Your Instinctive Mind is organized like a swarm of bees and it inhabits the billions of cells comprising your body. Your material body is organized out of the same chemicals as are in a shovelful of dirt, the only difference being that dirt is certain chemical substances in their crude state, while in tissue, bone and nerves, these chemicals have been acted upon by your Instinctive Mind, and refined into a suitable housing for your spiritual self. Imagine these tiny intelligent sparks using prana as the energy with which to transmute the crude chemicals taken in as food and water into flesh and blood, and realize the great importance of prana to the welfare of the body. There are a number of "storage batteries" such as the solar plexus in the body where a reserve of prana against emergencies is stored: usually in sickness these reserves are exhausted, but when you enter the silence they become "recharged."

Now consider "germs." Tiny sparks of malignant intelligence each housed in its little material body; they can only gain a foothold when an inharmonious condition exists in your Instinctive Mind. Otherwise, they stand no more chance of gaining a foothold than a hornet would if it entered a hive of bees. Even when germs do gain entrance, your Instinctive Mind has the power to organize a part of itself into antibodies which can annihilate the invaders.

Realize that each incident and misfortune of life is designed to force you upward on the ladder of spiritual evolution, and that even as astute and unscrupulous rulers quell inharmony among their own people by fomenting a war with another power, your Instinctive Mind, when inharmony impairs its efficiency, automatically creates the conditions which enable the germs to gain a foothold. It does not do this deliberately of course, but the inexorable Laws of the Absolute make it so to force harmony and happiness upon you. An habitual "worrier" is in effect extending the open hand of hospitality to any and all germs which may be seeking a home.

You know that regardless of the material shapes assumed by germs, or the shapes of the antibodies which fight them, the war is not between little microscopic natural bodies but between the opposing sparks of intelligence which activate them; and that from the storehouse of your Instinctive Mind, you can, if you will, hurl antibodies without number against the invaders. All serums used by the medical profession are designed to force your Instinctive Mind into forming antibodies with or without the cooperation of your Intellectual Mind, but as you gain mastery over your instinctive faculties, you will be able to do this at will and make yourself immune to almost every disease.

Realize that the only power capable of permanently influencing your Instinctive Mind resides in your own Spiritual and Intellectual Mind; it is the thoughts of inharmony and worry dropped into the Instinctive which keep it in tension, mobilized to fight *imaginary* bitter enemies, and thus preventing it from using its resources to repair damage or repel invading germs within.

Meditate upon the fact that you inherit the creative ability of the Divine Creator, and that the flowering of this ability will settle your every problem; it cannot flower in the arid land of fear and inharmony, because it requires the fertility of glad confident expectation.

Fifth lesson: enter the silence for fifteen minutes, then study the chapters: "Force and Resistance," "Desire," "Silence," and "The Emotions."

Now meditate upon the significance of "Force and Resistance." You may have often wondered why some "unfortunate" people seem to be continually beset by difficulties and misfortunes; yet if every difficulty could be solved for them and for you, and you were allowed to spend the remainder of life like a contented ox in a green pasture, you would rapidly degenerate to the spiritual and mental level of—a contented ox, and the upward march of evolution would be arrested.

Suppose that after teaching pupils the most elementary lessons a teacher should say:

"My children, I desire to see you happy, and I realize the strain and worry these lessons cause you; therefore, instead of teaching you the higher and more difficult lessons, I am going to allow you to spend the time enjoying yourselves on the playground." Would that not be absurd?

Like a child suffering from a headache over an examination paper, you are suffering from yourself—not the lesson. If the lesson is hard, you must master it by being hard with the lesson—the greater the resistance opposed to you, the greater the force you must employ in overcoming it.

Here is a little story to meditate upon.

Once, in a puddle of mud, a tadpole was hatched, and for a time he lived quite happily, wriggling and skittering through his soupy habitat by virtue of his long efficient tail. Gradually, he became aware of a terrible change taking place within his body; his beautiful tail was fast withering away and a pair of long ungainly legs were developing. Thus equipped he found great difficulty in moving around, and as the summer sun began to evaporate the water of his puddle and the mud became thicker and thicker, life became almost intolerable, and the effort expended in forcing his way about in search of insects for food left him exhausted.

Utterly discouraged at last, he decided to make one more trip to the surface, then to sink to the bottom of the mud and hibernate, until the autumn rains should again liquefy it enough to make life livable. While at the surface he took one last look around at the sweet green grasses of the bank, and the waving trees and he gazed enviously at a rabbit hopping by.

"Would that I had his legs," he thought, "then instead of living in a puddle, I too might bask in the warm sunshine and travel where I wished."

Then an idea came to him. "Perhaps even with the ungainly legs I possess I might be able to make my way slowly on dry land," so acting upon the thought, he hauled himself laboriously to the bank and away from the edge of the puddle.

How good was the sunshine! He flung back his head and croaked in pure rapture. Alas though, the sun's rays hardened the coating of mud which clung to him into a solid armor, and try as he would he could neither go back to the safety of his puddle nor ahead to the shelter of the bushes.

Then the shadow of a hawk fell upon the ground beside him and stark terror swept through him. Gathering himself together under its influence he gave a frantic leap toward the bushes, and the effort burst asunder the coating of dry mud which had bound him. Forever afterwards he was free to go where he wished through the sweet dewy grasses in long exhilarating leaps.

Now you see that Nature was not unkind in causing his tadpole tail to wither away in favor of a pair of powerful jumping legs, even though those legs could not do him much good in the mud. Neither was she malignant in drying up his puddle, for by making the old environment untenable, she forced him into a better one and a wider freedom than he otherwise possibly could have known. Even the prowling hawk did him a favor in disguise, by arousing the terror in him which burst his bonds.

So it is with our life's experiences—each one of us, every minute, is exactly where he should be for his own higher development. When we are beset with great difficulties we are face to face with opportunities, and life never gives us difficulties too great to overcome, if we can draw upon our higher powers to aid us.

You have learned that our desires are the driving force within us, and if we appear to be standing still in life, it is usually because we are entertaining two or more opposing desires, which pull us in two directions simultaneously. Make a list of your major desires and see how many of them conflict—you may be surprised—then make your choice between them deliberately and decide to discard all those least useful to you. Talk over the ones you would eliminate with a friend, if you wish, in line with the principle you learned in the chapter on science; then keep silent about the ones you wish to become realities. Draw them from your memory daily during your periods of meditation and fertilize them with your imagination, resting sure in the knowledge that by doing so you are watering the seeds and in due course they will sprout into manifestation.

The chapter, "The Emotions," has given you the method for withering the conflicting desires—examine them coldly and objectively whenever they intrude themselves into your mind, without allowing them to stir the emotions, then toss them aside and take from memory the opposite desires you wish to have realized and review them with all the imagination and emotion you can muster.

For this *last lesson*, enter the silence for fifteen minutes, then study the chapters: "Suggestion," "Removal of Limitations" and "Psychic Influence."

Now meditate upon the fact that our apparently colorless sunlight contains all the colors of the spectrum, and that in a garden ablaze with a variety of flowers, we see this sunlight split into an infinity of lovely colors and shades, by the pigments each flower has selected for its petals. This pigment allows certain rays of the sun to pass through the petals unimpeded, while turning back or reflecting other rays. In this way each plant "tunes in" on the rays most beneficial to it and discards all others.

In the same manner, the harmony of life is made up of an infinite number of blending notes, from thunderous bass to piping treble, and we can attune ourselves to its whole melody, or to one tiny part of it, as we will, and we shall reflect back, as the flowers reflect sunlight, whichever part to which we attune ourselves.

No event in itself has the power to materially alter the harmony of life—in fact the occasional jarring crash of the cymbals or roll of the drum, while in themselves anything but beautiful, are necessary to lend zest and variation to the theme. How absurd then, to attune ourselves to the cymbals and drum only, and to filter out the real beauty of the other notes.

Make up your mind that henceforth you are not going to allow the suggestion of others to rule your life—that when a person either wildly praises or bitterly denounces anyone or anything, he is just reflecting back an incomplete part of the theme, and when you are swayed by his suggestion, you are attuning yourself to be dragged hither and yon behind the wild horses of emotion without reason.

In the chapter "Removal of Limitations" you were given a picture of how the Instinctive Mind, in avoiding mental pain or disappointment, often prevents you from accomplishing a desired objective by withholding the very abilities necessary to its accomplishment.

You must eliminate the fear of disappointment and realize that the sweetest fruit of life is the pleasure of anticipation; that it is quite absurd to deny yourself this greater pleasure on the grounds that later you might be disappointed in not receiving the lesser pleasure of actual accomplishment. Imagine a gardener refusing to water his seeds because of fear that the plants may be blighted before they reach maturity.

Truth is endless and it is relative—a million photographs might be taken of a mountain from different vantage points and under different lighting conditions, yet though they were all true pictures, no two would be the same. By going to the silence, you attune yourself to all of those pictures and receive truth as a whole. If you develop the habit of doing this you will cease to be bothered by the occasional picture which offends you and the turbulence and anxiety will vanish from your day.

25

The Evolution of Consciousness

"The difference between a God and a worm is but the difference in the degree of their consciousness. A worm, having the illuminating consciousness of a God, would be a God; while a God, handicapped by the lack of consciousness of a worm, would to all intents and purposes be a worm."

Book of Right Feeling

You have been told in a previous chapter that *desire* is the propelling power behind all action; now we shall in turn, look behind desire and find that the compressed spring which pushes *it* into the forefront of consciousness, is the thing called *value*.

You cannot *desire* a thing unless you *value* it, yet if you *value* it you automatically *desire* it—but you cannot *value* a thing without believing that in some manner it can contribute to your happiness.

Every normal individual has within him the primal urge to *become more valuable to others;* that is the insatiable, nagging urge of creation which has goaded us upward from the caves to the mountain-peaks of knowledge where we can talk to the lightning and fly with the birds of the air.

Those outstanding people who are the leaders of the race in their respective fields, have succeeded in making themselves *more valuable* to their fellows then ordinary mortals. There is no doubt of that fact. Is this increased value manifested as some peculiar trait of character or personality? If we can answer that, we might have a clue to the mystery of personal value, which assuredly is not merely the old hackneyed "honesty, integrity, perseverance, etc." We all know dozens of excellent people who possess all of those qualities, yet who fail to become outstanding in any field of endeavor. What is this "super quality" which gives some a high value in the eyes of their fellows?

The answer is: *It is actually an increase in consciousness.* Often it is an increase in the quantity of consciousness, but always it is an increase in the *quality* of the consciousness as well. It is as though these fortunate people have the ability to view problems and events from a higher vantage point; they can see over the heads of the crowd and translate what they see into action.

Just pause here for a minute and consider your friends and acquaintances. Those you place the highest value upon are the keen, alert people, who seem immensely alive; the people who can by their swift understanding and sympathy, their flashing humor and brilliance, lift you while in their company to a higher level of *consciousness.* Those you value least are the dull, torpid, listless type who are slow to understand and to react. By the same token the first type are the most successful people—because they are the more *conscious.*

Consciousness also plays a decisive role in the maintenance of health, for it is an established medical fact that every change in consciousness brings with it physical repercussion, even as every physical disease brings with it a change in consciousness.

Physical illness often causes delirium, forgetfulness, hallucinations and even a complete loss of consciousness; states of consciousness, such as the anxiety state, may produce peptic ulcers, hyperacidity, diabetes, etc. In fact every change in consciousness, mild or severe, has a corresponding repercussion in every tissue and cell of the body.

The bearing that your consciousness has upon your health, however, is not limited to its immediate repercussions upon the cells and tissues of your body; there is also a secondary effect which might be illustrated as follows.

Suppose that you were forced to make a trip on a dark night, over a winding road in strange territory; your car had defective brakes and would only run at a high rate of speed; and your headlights were very dim. You would hang over the wheel tensely, peering through the windshield, with every nerve and faculty tensed to the highest pitch, to make such swift changes in your car's direction as were necessary to keep it on the road.

During that drive your total consciousness would be fully occupied and alert on the task of driving the car and there would be little of it spared to carry out the essential work elsewhere within the body, for the consciousness of the body tends to focus the body's energy upon the object of its attention, and so it would be partially withdrawn from the work of transmuting food, air and water into the hundreds of special hormones, enzymes, etc. necessary to the maintenance of the body's balance of health. There would be still less energy available to wage war against invading germs and to repair structures damaged in the regular wear and tear of living.

Suppose now that you were able to adjust a fuse, and then your headlights leaped into full brilliance, flinging their broad clear illumination for a mile in advance of the car—you would be able to relax and even to enjoy the drive, and the vital processes of your body would have a chance to use the energy released by your relaxation.

Well, just think of that twisting, unknown, dark road as the road of life, your car as yourself and the headlights as your consciousness. In the streamlined, industrialized, mechanized age in which we are living, we are forced, whether we will or not, to travel at high speed. In order to earn our living we must adjust our speed to that of the "caravan" regardless of our "headlights" or consciousness.

So the tenseness of our "driving" depends upon the dimness or brilliance of our "headlights," or consciousness, and since we cannot escape from the "caravan," our only hope is to increase the illuminating power of our consciousness or to suffer a breakdown somewhere in the physical machine that has not been allowed the time and energy necessary to repair itself.

However, by the time a breakdown does occur, the damage is often so extensive that repair is a long and tedious job, and to make it even more difficult, many patients continue to keep their consciousness tensed with worry and anxiety even after sickness has forced them to take time out for repairs. That is one of the reasons why so many patients fail to respond to medical treatment and either become "chronics" or die.

Keeping the foregoing facts in mind then, our problem is to first, *improve the quality and depth of our consciousness,* and second, *to learn to detach it from unprofitable contemplation at will.* In order to do this we shall have to briefly sketch the evolution of consciousness from a different point of view from that of the previous chapters on the development of various mental phases. We shall consider it now from the *pleasure/pain* angle.

The lowest forms of animal life are purely instinctive so far as their consciousness is concerned; they respond to stimuli from without *automatically,* even as they respond to the stimulus of hunger from within automatically.

The stimuli, both from within and without, which bring a response from these lower forms of life, and cause them to *act,* might be broadly classified as *painful.* Hunger is a form of pain, and so are most of the contacts from without which cause the forms of life in question to shrink back or flee. So we might say that all *action,* or rather all *individual action* in the lower forms of animal life, *begins as an escape from pain or discomfort.*

Now enters the sensation of *pleasure.* The act of eating is a *pleasurable* action, as opposed to the pain of hunger; the sensation of security after escaping from a painful or dangerous encounter is also a *pleasurable* one, and so it is the memories, in terms of pleasure and pain, attached to aspects of its environment, which slowly built up into an Instinctive Mind with all animals.

Pushed by pain and lured by pleasure, life climbed slowly upward, and ever as it climbed, it became less instinctive and more *conscious*. That stage of consciousness enjoyed by the higher animals today, and from which the human species has recently emerged, is called *simple consciousness*. Simple consciousness implies only awareness of the physical self and its physical environment. Creatures, at this level of consciousness, react on a purely instinctive level.

With the growth of instinctive knowledge, however, the sensations of pleasure and pain came more and more into conflict with each other; the greater pleasures were often guarded by a rampart of pain which had to be stormed before they could be enjoyed.

Thus, judgment and forethought were born and the development of the Intellectual Mind began. Still pushed by pain and pulled by pleasure through the thousands of years which followed, as the Intellectual Mind gained in power, humanity gradually entered into a new phase of consciousness.

This new phase of consciousness is called *self-consciousness*, because it carried with it a consciousness of the *self* as a being apart from every other thing in the universe. Humans still kept *simple consciousness*, of course, and in the early stages of Intellectual development life was still ruled by it; reason, or intellect, could only guide actions when those decisions it made were in full agreement with the pleasure/pain memories of the Instinctive Mind. Whenever the Instinctive Mind and the Intellectual Mind were in conflict, the Instinctive Mind always won—*as it does even to this day with the vast majority of people.*

However, this puny infant Intellectual Mind enabled the individual to step aside from the self and to ask: "To whom is this idea presented?" Or, to think: "The thought that I had about that thing was true; I know it is true and I know that I know it is true."

By self-consciousness the individual was also enabled to

disassociate from emotions and moods, and could learn to view himself realistically as a spiritual entity, having a mind and a body, instead of a body having perhaps a mind and an indefinite something called a "soul."

With the deepening of self-consciousness, there comes into being a still higher form of consciousness—*cosmic consciousness*, or *spiritual consciousness*, which we described briefly in a previous chapter. *However, as this form of consciousness is the goal to which humanity has been pushed by pain and pulled by pleasure for at least half a million years, a further explanation of it will not be amiss.*

Spiritual consciousness is as far above self-consciousness as self-consciousness is above simple consciousness, and even as the Intellectual Mind's development lifted humanity above the ranks of the lower animals, spiritual consciousness lifts its possessors higher above the ranks of the average person, though just as we retained simple consciousness with the development of self-consciousness, the individual who passes to spiritual consciousness retains both other phases of consciousness as well.

Spiritual consciousness brings with it an illuminating understanding of the life and order of the universe, and a brilliant intellectual enlightenment which makes even the most complex problem of life seem simple and clear; coupled to these is a sense of lofty moral elevation and a sense of exaltation and joyousness impossible to describe. There comes also a conviction of immortality—not a feeling that one shall have it, *but a certainty that one already has it.*

As I explained in a previous chapter, even a faint glow of spiritual consciousness tranquillizes and harmonizes every function and process of the body, and sometimes even speeds the normal healing processes into a triumphant surge which almost instantaneously obliterates every trace of disease. Every "miracle cure" from the dawn of time has been accomplished by the patient getting a flash of spiritual consciousness, even as every religion was started by some enlightened person attempting to lead followers into this phase of consciousness; the Kingdom of the Christians, Nirvana of the Buddhists, and Tao of the Taoists does not describe a place located somewhere behind the skies, but is merely a different name for spiritual consciousness.

As might be expected, there are many different degrees of spiritual consciousness, even as there are many degrees of self-consciousness, and it often manifests first in the field of the possessor's greatest interest; thus, we might see a mathematician operating in that field with a very high degree of spiritual consciousness, while remaining on the plane of self-consciousness in other fields in which he is less interested, or we might see a person operating in the mechanical and organizing field in the spiritual consciousness, while perhaps even retaining the naivete of simple consciousness in another field, such as politics. However, just as the dawn lights up the mountain peaks before it floods into the valleys, it is just a question of time until the dawning spiritual consciousness of the possessor floods into every aspect of life.

Every great advance in thought, every great musical composition, every great work of art or literature is the product of its author's spiritual mind, for that is the *only* source of inspiration. Even if we accept the possibility that certain individuals throughout history received their inspiration directly from the Absolute, it would still have to come by way of the spiritual consciousness.

Now let us summarize briefly what we have said so far in this chapter before passing to a discussion of ways and means of practical application.

We learned that the sensations of pain and pleasure were stored memories of our simple consciousness about conditions in our primitive past; gradually, as our simple consciousness developed, we learned to seek those things which gave us pleasure and to avoid those things which gave us pain. In this way we were kept on our narrow evolutionary pathway.

With the unfolding of our self-consciousness, however, a

new element entered into the situation—the element of *duty, the knowledge of right and wrong.* Often, in fact very often, the voice of our ancient, primitive simple consciousness goaded us into a course of action that our self-consciousness told us was wrong and discreditable; if we yielded to the lower urge we paid a high price in remorse. *So the pleasure enjoyed by our simple consciousness often became pain to our self-consciousness.*

In the above stage of evolution (and most of us are in that stage today) we are as a house divided. If we obey our simple consciousness, we pay a terrific penalty of remorse and self-condemnation; if we obey our self-consciousness, a condition of emotional torment results which often ends in a neurosis or psychosis.

In fact with the development of the Intellectual Mind and the unfolding of self-consciousness, we are placed in an intolerable situation; if we yield to the voice of our Instinctive Mind, or simple consciousness, often the results are intellectual pain and a sense of guilt, yet if we fight that primitive voice within, a serious emotional upset might result. Instead of a straight pathway, with pain behind and pleasure beckoning, we are in an impasse, with pain glaring from in front and behind. Is it any wonder that a twenty-fourth part of our entire population enters mental hospitals for treatment, that there is a steadily mounting toll of nervous ailments and that functional diseases of mental origin are on the increase? Humanity is in a stage of transition and most of us are hovering between simple consciousness that is without "knowledge of sin" and spiritual consciousness, which is "beyond sin."

Nature never places her children in an intolerable position without providing them with a means of escape, and she has done just that in the present situation. We cannot escape from our self-consciousness by going back into the egg of simple consciousness from which we hatched, *but we can escape by going forward into a higher degree of consciousness.*

The first step is to deliberately determine upon a change of masters. We are going to change our old pleasure-seeking, pain-avoiding ways of life. We are going to break the chains of pleasure and pain that have shackled us to a primitive dictator within ourselves for half a million years; *we are going to be free as individuals to serve the cause of right, regardless of pleasure or pain.*

The second step is to realize that hitherto we have suffered from spiritual amnesia. Amnesia, is a condition in which the sufferer forgets who he is; he may be quite lucid and intelligent, but he has a complete lapse of memory of his real identity. Well, most of us have forgotten who we are and what we are in a spiritual sense, and we are going to have to develop the memory before our spiritual consciousness can awaken.

In order to develop this memory, we are going to have to superimpose a new thought pattern over an old one. A thought pattern is something like a recording in the brain which plays back the same fixed idea in response to a set of opposite ideas. We have accepted for so long that we are merely physical bodies with minds, that this idea is graven in our brains, almost as deeply as the idea held for centuries that the world was flat. Even as that absurd idea was overcome in the minds of most people by a continual repetition of the truth that the world is spherical, so the equally absurd idea that we are merely physical bodies with minds can be overcome by our constantly repeating to ourselves the truth that we are spiritual entities, who have woven bodies.

It is a good plan to set aside a study period every day, and during that period to refuse to be disturbed. Go to your room and lock your door at this time; make it clearly understood that for the hour you are to be considered as inaccessible as though you had gone for a walk. Read over one chapter of this book when you have settled yourself comfortably, beginning with the chapter on "Imagination" and proceeding with another chapter each night, until you have reached the end of the book; then begin again as before. After slowly and thoughtfully reading the chapter, do the exercises outlined in that particular chapter, faithfully and thoroughly, then finish by entering the silence, as explained in the chapter "Knowledge Harnessed."

It is a good plan to start keeping a diary. Divide each day's space by two, so that you may make one entry in the morning upon awakening and another before retiring for the night. Each morning, enter in it briefly what you plan to accomplish in the way of eliminating the pleasure/pain tyranny of your Instinctive Mind. It might be that you will decide to do one thing each day that you have hitherto disliked doing, then at night make a record in your diary of your success or failure. In the course of a year, this diary will make interesting—and inspiring—reading.

Plan to avoid hasty decisions. Upon taking your problems to your study period, where you should review all the elements as calmly and judicially as though the problem belonged to another person, go into the silence and "listen" for an answer. Write this answer down in detail, then study and analyze it for the bearing either pleasure or pain has had upon its formation. If you find that it is the product of either the fear of pain or the desire for pleasure, tear it up and try again. *Continue this until you are satisfied that the answer is as impersonal and just as though it came from the Absolute.*

Form the habit of thinking of your Intellect as the center of a pair of balance scales; on one side is the ancient and highly developed Instinctive Mind, and on the other is the young Spiritual Mind. Each idea that you entertain is a grain of sand that you must place on one side or the other. At present the Instinctive Mind side of the scales heavily outbalances the Spiritual side, for you have been heaping grains of sand on it for countless centuries, so you will have to work hard to counterbalance the Spiritual side, but every grain of sand placed there adds to that new balance of life.

You must understand that the *motive* attached to each idea is what determines the side of the scale on which it belongs. If the motive is pleasure or pain, it belongs to the Instinctive side; if it springs from the desire to, as a free Spiritual individual, leave this world a little less selfish, a little more tolerant, a little less harsh and little more beautiful than you found it, regardless of either your personal loss or gain, then it belongs on the Spiritual side.

Gradually, as you practice this giving up without reserve all longings and desires arising from the pleasure/pain motivation of the simple consciousness, you will find a new and higher motivation replacing it—the motivation of the Spiritual Mind, and then you will understand the meaning of the word *freedom.*

"But," I hear you say, "am I to disregard my own interests utterly, and in many instances even suffer grave losses through practicing such a philosophy?"

The answer is, that by escaping from the thraldom of the simple consciousness, and by developing the spiritual consciousness, you are giving up the philosophy of "the over-provident dog who buries bones in the wake of the caravan," and you are placing yourself in tune with the creative might of all the universe. Your every plan, if founded upon these higher ideals, *will have a million unseen allies working with it, instead of against it.*

My wife and I, both trained in the technique of scientific investigation, entered upon a series of experiments which stretched over a period of more than a dozen years, to test the principle here outlined. Time after time grave situations developed, when the *right* thing, judging by all the accepted standards, seemed the unwise thing to do, yet time after time with the unfailing regularity of any other law of the universe, the *right* thing brought results which were almost incredible and success beyond our hopes.

So even if one were unintelligent enough to reject the universal and intuitive belief, held by all races and all religions from the dawn of intellectual development, that "sin" is penalized, and merely wished to increase his potentialities as a human being, and his potency in his chosen work, the foregoing would still be the best and shortest method of achievement, for sin or evil is thought which springs from the narrow, personal pleasure/pain motives of the simple consciousness or Instinctive Mind; good or right is thought and actions which aim at being *right* regardless of either pleasure or pain.

Lastly, all religions agree that those who have achieved spiritual consciousness have banished death and have won individual immortality; that when it comes time to doff that "fleshy overcoat" we call a body, we pass in full consciousness to a higher, brighter field of endeavor and *only death dies with the body.*

26

The Adventure of Life

"Never the spirit was born—the spirit shall cease to be—never; Never was time it was not—ends and beginnings are dreams; Birthless and deathless and changeless remaineth the spirit forever; Death hath not touched it at all, dead though the house of it seems."

Bhagavad Gita

In upper India I once had the privilege of meeting a most unusual old man; the sun was scarcely an hour high when I came upon his flower-embowered retreat. From the snow-capped Himalayas in the background to the mist-enshrouded valley below, the world seemed a dream of fresh loveliness. In this setting he stepped forth to meet me, an erect, powerful figure, standing firmly on his sturdy legs, head flung back and a welcoming smile on his lips.

I gasped in amazement, for in the village below they told me that this man had lived for a century, yet he looked like a particularly strong man in his prime.

Later, in the course of conversation, he explained his formula for the preservation of youth.

"Nature," he said, "abhors monotony—step outside and examine the view one hundred times a day, and you will find it different each time. The effect of light and shadow is the principal cause of this. Yet, not content with this hourly change of scenery, Mother Nature has a complete change of dress several times each year; the snows of winter give place to the tender green of spring, and that to the browns and orchres of summer, which in turn yield to the grays of autumn.

"We live, doctor sahib, not in a fixed order of things, but in a world of continual flux and change—old nations and civilizations wane and die, and upon their ruins are built new and different civilizations—social systems which seem the very soul of permanence and serenity are swept away overnight and new systems are installed. A single scientific discovery can alter the political and economic destiny of the entire world. Why, those very mountains—the backbone of the world—are being steadily broken down by the forces of erosion and their particles carried away by the rains and rivers to form new lands elsewhere.

"You doctors tell us that the very cells of which our physical bodies are formed are dying and new ones being born every second, so that from the marrow of our bones to our hair, we change our bodies completely many times over the years.

"Is it not wonderful to think of the human spirit, the Soul, breasting its way through this great changing sea of events, like a shining light high on the mast of a staunch vessel, the only changeless thing in an unstable world, unfolding more and more of light-giving potentialities through contacting new situations and problems?

"For this changing sea was designed by the Creator as a school for the soul, Sahib. The only way in which the soul may unfold its infinite potentialities is through experiences. This then, is the secret of life and happiness: seek new experiences —disdain stagnation—each day, seek to go forward spiritually to new and higher adventures. The body does not die as long as the Soul needs it. It is only when we have ceased to learn, or become bored by monotony that we lay aside our bodies, and, after a rest to digest our experiences, come back in a fresh, new body in a new environment for new experiences.

"When I was seventy I learned something about painting

pictures. Five years later, I took up the violin. Since then, I have taken up horticulture, learned several new languages, and am at present studying astronomy."

"But," I interjected, "some would say that the price of constant study is too high a one to pay for a mere continuance of physical existence."

The old man smiled. "And they would be correct. It would be too high a price for mere physical existence, but by study we learn more about the great universe of Brahma, and through knowing more of it, we are better equipped to enjoy it and to seek new experiences in it, and thus unfold more and more of our soul's inner nature. You see, no knowledge is ever lost. We have all eternity before us and what we learn in this life, we shall possess in the form of instinct in the next. Have you not seen youngsters even in your country, prodigies I think you call them, born with certain forms of knowledge and at an early age able to far out-distance elderly experts in the same line of knowledge?"

The old Guru was wise and his formula a good one. Stagnation spells death in any language. We are all, whether we will it or not, spiritual adventurers forever advancing to that mystic land behind the ranges where beauty shall reign and the rough pebble of the soul shall be cut and polished on the wheel of circumstance and experience until it reflects that beauty with a joy unutterable.

Brutality and injustice inflicted upon us when we are powerless to retaliate—the treachery of one whom we loved and trusted, sickness which holds the body prisoner and tortures it upon the rack of pain—these and many other experiences seem cruel, but would you be willing to part with any experience which you have ever undergone? I think not. All of us, perhaps, would like to eliminate from our lives some painful memory of a disgraceful episode, but we would not be willing to part with the knowledge gained from it, for we would be parting with a little bit of our character—ourselves. If we continued to part with the fruits of our experiences, we would inevitably return to the prehistoric state, for all knowledge is but recorded experience.

Just as a gem cutter turns his precious gem this way and that way against the polishing wheel in order to wear away the roughness and cut the facets which will show the stone's true brilliance and beauty, so Universal Mind polishes our soul upon the wheels of experience, and from each one we gain a little more inner understanding. Each of us, every moment, is exactly where he should be in order to acquire the next necessary experience. We have, during past lives, undergone many experiences and in the future we shall undergo many more. All of us know that certain things are no temptation to us and we wonder at our friends being tempted by them. But that merely means that we have undergone that experience in some past life and have no longer any need for it. Thus, it ill-becomes us to be self-righteous or critical of anyone, for even the thief, the prostitute and the murderer are merely in a different grade of the same school as ourselves—our Alma Mater—life, and they are in a grade we have either graduated from or have yet to pass through. Furthermore, remember that when you find yourself criticizing another, you are touching a chain of cause and effect which must inevitably make you yourself the object of criticism, and that in due course the spotlight you turned on that person will be turned upon you with interest.

You have gained much through your illness; you have gained tolerance toward the complaints of others, sympathy with suffering, pity for those whose lives have apparently been frustrated, patience and courage with your own complaint, and you are spiritually richer through having undergone the experience. But be careful lest pity and sympathy entangle you in a morbid morass of futility. Remember always that *you* stand alone. *You* are passing through this changing sea of life alone and, although you may meet many delightful friends and perhaps bitter enemies, they are like ships that pass in the night—they join you for a moment in a blaze of light, then pass again into the darkness. They also have a destination, and are proceeding inexorably toward it. Even loved ones, whom we would keep forever, must pass from our picture as perhaps many of you have experienced in the death of some member of your family. The course of wisdom, then, is to enjoy our friends to the utmost, as we would fellow travelers in a caravan, or as interesting fellow passengers on a great ocean liner, helping them all we can, but realizing that inevitably the caravan must reach its destination—the liner must dock and, with the reason for the association gone, we must say *au revoir*—to meet again perhaps in this life or another.

Learn not to expect help from another, but to look for it within yourself, for *you* possess all of the powers and attributes of every soul alive, and your intellectual mind creates the demand for their unfoldment. When you depend upon another to solve your problems, you are passing up an experience and retarding your own development. Better by far to try, and fail, yourself, than to leave your problem to someone else to successfully solve. For it is through our failures that we learn sometimes more than by our successes.

It might be summed up in the old Eastern saying "No man is your friend—no man is your enemy—but all alike are your teachers." If you cultivate the scientific attitude toward life, condemning nothing, expecting nothing from others, but doing your utmost for all, not for praise or in expectation of reward, but because in doing so you are intelligently cooperating with the plan of the Absolute and fulfilling your destiny, you will escape from monotony and futility and discover that as you use your own power, more shall be given you.

Lastly, remember that *power* is not the child of weakness, nor is helpfulness the offspring of sloppy sentimentality. Better by far to pull a faltering comrade to his feet and urge him forward again with a battle cry than to fall upon his neck and shed tears of sympathy. The tools you have to work with are force and resistance, and those philosophies which weaken spiritual force and spiritual resistance are useless philosophies, as the laboratory of history conclusively proves. The only devil in existence is the devil of inertia, who is ever-present to whisper seductively to choose the easier path, or to do nothing or to let someone else do it, until the victim becomes a flabby caricature—the weak echo and tool of any stronger character. When whole nations become thus afflicted, they are ripe for invasion by the lean, hard barbarians, who take away liberty that has been squandered, until, perhaps through centuries of slavery, the spiritual fat is worn off and a leader is born among them who is capable of giving them a new trial at liberty.

In the battle of life, learn to be a happy warrior, looking neither to the dead past nor to the unborn future. *Now* is all you have. If you could fly forward with the speed of sound, and someone shouted "Now" that sound would be ever with you as you flew onward—you would live in the eternal *now*. Well, you are in eternity *now* as much as you shall ever be and *now*, this moment, and the next and the next you are creating your destiny, so do not say to yourself, "tomorrow, I shall begin practicing the facts outlined in *The Healing Power of Your Mind*." Begin *now* and enjoy the fruits in that which shall automatically replace "tomorrow."

I hope you have enjoyed reading this book as much as I have enjoyed writing it. I hope that the knowledge herein may help you recapture the priceless boon of health. But more, I hope that it will help you recapture the sparkle of sunlight, the colors of the sunset, the jeweled stars, the reflection of the Absolute in the face of humanity and the great living, breathing, beautiful earth. I hope that it will re-awaken your keenness in the greater adventure in the universe—the adventure of life—and that in your reflections at eventide, there will come the feeling that it has been a privilege to live through the joy of the day.